T0290844

Evolutionary Optimization Algorithms

Evolutionary Optimization Algorithms

Altaf Q. H. Badar

CRC Press is an imprint of the
Taylor & Francis Group, an **informa** business

MATLAB® is a trademark of The MathWorks, Inc. and is used with permission. The MathWorks does not warrant the accuracy of the text or exercises in this book. This book's use or discussion of MATLAB® software or related products does not constitute endorsement or sponsorship by The MathWorks of a particular pedagogical approach or particular use of the MATLAB® software.

First edition published 2022
by CRC Press
6000 Broken Sound Parkway NW, Suite 300, Boca Raton, FL 33487-2742

and by CRC Press
2 Park Square, Milton Park, Abingdon, Oxon, OX14 4RN

© 2022 Taylor & Francis Group, LLC

CRC Press is an imprint of Taylor & Francis Group, LLC

Reasonable efforts have been made to publish reliable data and information, but the author and publisher cannot assume responsibility for the validity of all materials or the consequences of their use. The authors and publishers have attempted to trace the copyright holders of all material reproduced in this publication and apologize to copyright holders if permission to publish in this form has not been obtained. If any copyright material has not been acknowledged please write and let us know so we may rectify in any future reprint.

Except as permitted under U.S. Copyright Law, no part of this book may be reprinted, reproduced, transmitted, or utilized in any form by any electronic, mechanical, or other means, now known or hereafter invented, including photocopying, microfilming, and recording, or in any information storage or retrieval system, without written permission from the publishers.

For permission to photocopy or use material electronically from this work, access www.copyright.com or contact the Copyright Clearance Center, Inc. (CCC), 222 Rosewood Drive, Danvers, MA 01923, 978-750-8400. For works that are not available on CCC please contact mpkbookspermissions@tandf.co.uk

Trademark notice: Product or corporate names may be trademarks or registered trademarks and are used only for identification and explanation without intent to infringe.

Library of Congress Cataloging-in-Publication Data

ISBN: 9780367750541 (hbk)
ISBN: 9781032073378 (pbk)
ISBN: 9781003206477 (ebk)

DOI: 10.1201/b22647

Typeset in Nimbus font
by KnowledgeWorks Global Ltd.

Dedication

My parents (in Heaven), my wife, Sajeda, my two lovely daughters, Manaal and Alfiya, along with the blessings of my Spiritual Masters and the Almighty

Contents

Preface

Artificial Intelligence (AI) has come into our lives a big way and it is here to stay. From the apps in our mobiles to the floor cleaning robots to the electronic assistants, it is everywhere. There are different branches of AI, like neural network, fuzzy logic, and more. One of the most important branches is the Evolutionary Optimization Algorithm (EOA). EOAs are known by many terms like *stochastic optimization method, swarm intelligence*, etc.

EOAs are descendents of conventional optimization methods like linear programming, quadratic programming, etc. However, there is a stark difference between them, as the conventional methods worked on a single solution and improved it, whereas the EOAs employ multiple solutions and try to find the optimal solutions. The conventional methods were quite time consuming and were complex to implement. The EOAs, as we will see in the book, are quite easy to understand and implement. Also, the EOAs can be easily applied to any optimization problem, or in other words, the EOAs are quite adaptable in nature. EOAs may not present the best/optimal solution but are capable enough to present near optimal solutions in a short time duration.

EOAs have their own share of problems, but these problems can be ignored in favor of the benefits that these algorithms bring to the solutions of the optimization problems. EOA has a major drawback in that it cannot be implemented without the help of computers. These algorithms work on multiple solutions simultaneously and hence, it is out of the question to solve these methods through hand calculations. There is another reason for the requirement of computers while solving EOAs and that is the generation of random numbers, which is required in all of them.

There are multiple EOAs present in the "research" world, however, some of them may be quite similar in certain ways and a world apart in other ways. Some of the most researched EOAs are the genetic algorithm (GA), differential evolution (DE), particle swarm optimization (PSO), etc. This book has covered all the EOAs that have been in the top research fields and applications.

Areas like economics, engineering, etc. are detailed in a large number of books for the undergraduate level. Books on AI are fewer in number at this level, especially EOA. This book aims to fill this gap. The book chapters and its contents have been arranged in such a way that it would be ideal for an introductory level. The book should also serve as a one-stop destination for understanding EOA. The book shall be ideal for first-time students of this subject, even at a master's level, or for those research scholars who are just starting their research.

The book chapters are arranged as: basic concept, terminology, process, algorithm, flowchart, and example. This common format should help students compare the EOAs and find similarities amongst them. The example section at the end of each chapter gives a step-by-step solution of the EOA under consideration for two iterations. The problem solution for each EOA starts with the same set of input variables and solves the same Greiwank optimization problem. This book should encourage faculties to introduce optimization techniques as a theory and lab subject in the curriculum of their respective universities.

The understanding of EOA is further enhanced in the Appendix section through the addition of its application for a real-time system. The Appendix section has three parts, with the first one showing the workings of a real-time application of PSO in MATLAB® software, the second part dealing with EOA in Python, and the last part containing a set of standard optimization problems. In the first part, a live demo is presented to create a real-time application to solve an optimization problem in which the parameters can be changed at run time. Python is a fast emerging programming language and has multiple packages dedicated to EOA. Lastly, at the end when you have become confident in the optimization techniques, you can try your hand at solving some of the standard optimization problems.

I have been working at the helm of implementing optimization techniques in various fields. This has been my research topic for a long time and thus it gave me inspiration to write a book at the introductory level, as I could not find a similar reference. I believe this book can serve as a reference for many budding researchers.

Happy reading and optimizing.

Altaf Q. H. Badar

1 Introduction

1.1 INTRODUCTION

Humans have the intelligence to understand, analyze, improve and learn from the processes happening around them. In animals, too, whether through application of knowledge or unknowingly, the processes in which they are involved keep improving. The monkeys are known to develop tools. The crow has been trained to gather trash in return for food. Moths have developed techniques to evade bats over a period of time. A herd of elephants is able to find water during the draught season. Even trees in tropics grow high to beat competition and get some sunlight. The examples are many.

It is not a matter of intelligence but of behavior and evolution that the processes in which creatures 'having life' are involved will tend to improve over time. The processes are fine-tuned to return optimal methodology and results with the passage of time. The researchers and academicians have observed these behaviors of process optimization and emulated the methodologies involved to develop algorithms, which are termed, Evolutionary Optimization Algorithms (EOAs).

As has been mentioned, the processes get optimized by the persons/animals/creatures involved in performing them. Thus, it becomes very necessary to understand what optimization is!!!

Optimization is the process 'to optimize'. But then what is meant by 'optimize'? There are many aspects to the word optimize. It can be defined as 'to make something as good as can be', or 'to use something in the best possible way' or 'to make best possible use of available resources' or 'to improve the efficiency of something or make it work efficiently'. Let us concentrate on 'making best possible use of available resources'. One should concentrate on the fact that there is a constraint involved in the statement through the term 'available resources'. The definition also has another aspect, 'best possible', which means that it may not be the best; as for the given conditions, the best possible solution may not be implementable.

In the coming sections in this chapter, we will cover the formulation of an optimization problem, a brief review of the optimization techniques that are covered later in this book, and a wrap-up through the conclusion.

Before moving on, remember, "When all you have is a hammer, everything looks like a nail". In other words, optimization techniques have specific applications and cannot be applied for every problem. One needs to identify the problem as an optimization problem, otherwise there are many other artificial intelligence techniques which have specific specialization.

1.2 TERMINOLOGY

In this section, a brief terminology is presented for optimization techniques. This will help readers get acquainted with the optimization techniques. The same terminology

DOI: 10.1201/b22647-1

is used for all optimization techniques throughout the remaining chapters. To look at it from the other side, all evolutionary optimization techniques have some common features associated with them which are listed below:

Input Variable: The optimization problem to be solved through the EOA is dependent on certain values which have to be input. These values are termed input variables. The range of these variables are specified and a certain combination of these variables will generate the optimal solution. A combination of the input variables will represent a potential solution. Lastly, the random solutions generated at the beginning of an evolutionary optimization algorithm shall have these input variables generated randomly. For example: $y = 5x_1 + exp(x_2) + x_3^2$ has x_1, x_2 and x_3 as input variables.

Individual/Solution: EOAs require input variables whose combinations are to be optimized. An individual/solution shall have one representation of each of the input variables. In other words, an individual/solution also represents a potential solution to the optimization problem. While programming, the individual/solution can be represented through an array. For example: [3, 5, -1] as values for x_1, x_2 and x_3 is an individual solution.

Population: A collection of individuals/solutions is termed a population.

For example: $\begin{bmatrix} 3 & 5 & -1 \\ 0 & 2 & 5 \\ -1 & -2 & 1 \end{bmatrix}$

where each row represents one individual solution and its collection is termed a population. It is usually stored as a matrix.

Objective Function: The mathematical representation of the optimization problem at hand is termed an objective function. The objective function will be a function of input variables, directly or indirectly. An example is given in Input Variable.

Objective Function Value: When the input variable values stored in an individual/solution are fed in the objective function of the optimization problem, we obtain the objective function value. In other words, it is the evaluation of objective function by entering values in it.

Limits: Limits are the boundaries that define the range of an input variable. The limits can be minimum limit below which the input variable cannot have a value and maximum limit above which the input variable cannot have a value.
For example: $-3 \leq x_1 \leq 5$
which means that the value of x_1 cannot be less than -3 or greater than 5. These limits are helpful to define boundaries, in case, the search process violates them.
One last point—it is not compulsory that an input variable should have limits. In some cases, these limits may not be specified.

Constraints: Constraints are the restrictions that have to satisfy while solving an optimization problem. There are two types of constraints, which can be encountered while solving an optimization problem. In the first type, the constraints are to be verified before obtaining the objective function value of a solution, e.g., limits. Then there is a second category of constraints which are found after or during the evaluation of the objective function, e.g., voltage magnitude of a bus, state of charge of a battery, etc.

Search Space: The imaginary multidimension space defined by the limits of input variables is the search space. The number of dimensions shall be equal to the number of input variables. For a single variable, it would be an axis, for two variables it would be a plane defined by boundaries, for three variables it would be a box-type structure and so on. A simple search space is shown in Fig 1.1.

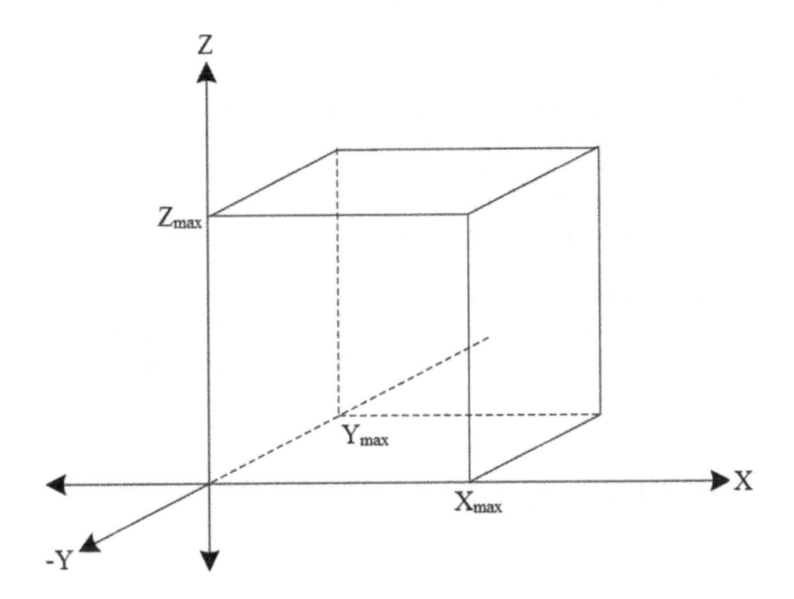

Figure 1.1: Search Space

In case the objective function is based on equations based on circles, cylinders, etc., the search space will be of similar appearance. However, this is applicable only up to three dimensions which can be visualized. For more than three dimensions it becomes imaginary.

Step Size: The input variables can be of two types: continuous or discrete. In continuous variables, any value can be assumed within the limits, whereas, in discrete variables, the step size is defined. Thus there is no step size for continuous variables, but should be defined for discrete variables. If a discrete variable is defined with limits of 0 to 10 with a step size of 2, then the variable can have values as 0, 2, 4, 6, 8 and 10 only. No other values would be allowed.

Iteration: Any EOA is able to find the optimal solution/result by repeating a certain sequence of steps. Once such cycle or set of sequence of steps/equations is called an iteration. The number or iterations to be executed shall depend on the optimization technique, programmer and the stopping criteria.

Stopping Criteria: The stopping criteria are the conditions when the optimization process shall stop. Based on various available references, there are three popular methods to stop the optimization process. First, the optimization process shall stop after a predefined number of iterations. Second, all the individuals have the same input variable combination; i.e., they converge. The third and last method implies that the optimization process shall stop if the best solution does not change for a specified number of cycles/iterations.

Local Optima: In a given search space, there may be solutions which may appear to be optimal, as there are no better solutions in the vicinity/neighborhood. However, they are not the optimal solution. Such solutions are termed Local Optima. Some of the other terms used for such solutions are local maxima and local minima, based on the type of optimization problem.

Maximization/Minimization Problem: In an optimization problem, there is an objective function. This function needs to be optimized under the given constraints. The optimization may be either maximization or minimization. Based on this classification, the optimization problem is defined as a maximization problem or minimization problem. These problems are interchangeable, and a minimization problem can be converted into a maximization problem and vice versa.

1.3 OPTIMIZATION PROBLEM

The optimization problem has been a common and continuous problem throughout time. All creatures, animals or plants keep optimizing so as to get optimal results. It's just that we never recognize this while solving such a problem. A simple example is visible through our travel, in which we try to choose the best route. How do we define a best route? It is the route which is shortest, fastest and least costly. It may be possible that we may not get the best in all the three factors; however, the output may be the most satisfying.

A very simple way to solve a problem is to divide it into three parts as shown in Fig 1.2. Any problem will have input(s), output(s) and a process to solve it. The process used to solve a problem is supposed to respect/satisfy the constraints. Thus, constraints act as a covering in the optimization process to stop them from crossing the specified limits.

The input values are fed into the optimization process. It should be noted that these input values should be within the specified limits, if these limits are specified. The input values are usually generated randomly for the first time and then calculated through the optimization process. The optimization method then puts these values into the process and obtains the output which is termed an objective function

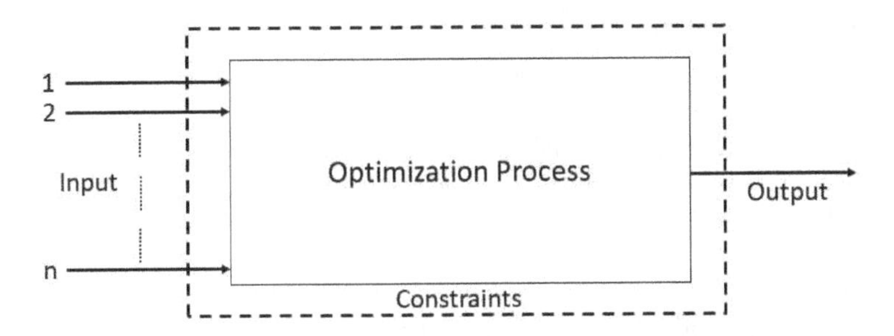

Figure 1.2: Optimization Problem Block Diagram

value or fitness value. The objective function has a single output or a combination of outputs (i.e., multiobjective problem) that we want to optimize. Optimization can be minimization or maximization. The objective function value obtained as an output of the optimization process is compared and analyzed by the optimization process for further changes in the input variable values.

Let us understand the process through a simple example: in electrical engineering, there is an optimization problem which requires reduction in the amount of reactive power flowing in the system. For a brief background of the problem: the power is generated through generators, then transmitted through transmission lines to the load. There are two types of power: active and reactive. The reactive power does not perform any work and need not be a part of the transmitted power as it will increase the losses. To control the flow of reactive power, there are certain parameters that can be varied, like generator bus voltages, transformer tap settings and use of capacitor banks.

In this case, the input variables would therefore be generator bus voltages, transformer tap settings and the reactive power from the capacitor banks. The number of input variables may sum up to a large value depending on the type of system the problem is dealing with. This optimization problem cannot be solved through normal methods as a large number of possibilities exists. To better understand, let's take an example of an IEEE standard six bus system as shown in Fig 1 (Appendix).

In this system there are six input variables: V_1, V_2, T_{43}, T_{56}, Q_4 and Q_6. An optimal combination of these input variables will minimize the flow of reactive power in the system. The next part of the problem is to identify the range of input variables and their step size; the details for the system being considered is given in Table 3 (Appendix). There would be very large number of combinations from which one combination would be optimal and is to be found. Hence, a normal method cannot be utilized to obtain the optimal state.

Now No we know that an optimization technique would be required to solve the problem, let's analyze it further. The ranges of the variables shall define the search space. A generalized method toward implementing the optimization technique would be as given in Algorithm 1.1.

Algorithm 1.1 Generalized Algorithm for Optimization Techniques

```
Initialize and Input required values
for COUNT = 1 to Pop do
   Generate random input variable values to form individuals
   Check that the generated values are within allowable limits
   Evaluate Objective Function value
end for
while Stopping Criterion not met do
   As per optimization process, analyze the population for best individuals
   Generate new solutions/input variable combinations
   Check and correct limits of generated individuals
   Evaluate objective function value of newly generated individuals
end while
Display Results
```

In any optimization technique, a random population is generated, such that the input variables should lie within the specified ranges. Next the objective function value for each individual within the population is evaluated based on their input variable values. It is necessary that the input variable values may lie within the specified range but the combination may not be feasible. Like in the above example, even though the input variables may be within the specified ranges, the combination should also satisfy load flow constraints. Constraints are covered in a later section. Right now we can say that the solutions should be feasible, satisfying all the constraints.

Once the objective function values for all the individuals have been evaluated, the optimization process is applied, through which new solutions are generated, i.e., new input variable combinations are generated. Based on the method being applied, the individuals making up the population shall change. The objective function value of these newly generated individuals is also evaluated. Again the optimization technique shall generate new solutions through its own process. The cycle shall keep continuing until the stopping criterion is met.

Some very common terms related to these optimization problems are individual best and global best. Individual best is a combination of input variables that give the most optimized solution from all the combinations that this particular individual has had through all the iterations of the optimization process. Global best is the combination of input variables that any individual in the population has had during the process of optimization until the current iteration.

1.3.1 CONSTRAINTS

Constraints are the limitations applied by the optimization problem. There are two types of constraints: direct and implied. The direct constraints are applicable during the generation of individuals. These may be used to define the boundaries of the search space and limit the input variable values within specified limits. The implied constraints are not applicable during the generation of individuals. Such constraints may be a part of the objective function value evaluation process. They may lead to some legal solutions being rendered as infeasible after evaluation of objective function value. An example of such an instance would be bus voltage limitations in the above discussed example of reactive power minimization in a power system. Even though the generator bus voltages are generated randomly as input variables, there are other buses present in the system as well. The limits of the voltages at these buses are also specified. Thus when an individual solution is generated, the input variables may be within the limits specified and thus legal. However, during evaluation of the objective function value, it becomes apparent that other bus voltages are not within limits. Then in such a case this individual solution is rendered illegal and has to be removed from the population.

1.4 MULTIOBJECTIVE OPTIMIZATION PROBLEM

In an optimization process, the objective plays a very important role. This objective of the process may be a single objective or can be multiple objectives. For example, reduction in cost represents a single objective; however, reduction in cost as well as time represent multiobjective problem. In some cases, the multiobjective problem can be formulated to form a single objective problem by applying weights and combining the objectives. In case the objectives cannot be combined, some variations of optimization techniques are applied like nondominated sorting genetic algorithm. In multiobjective optimization problems, a pareto front is formed which helps in solving the problem as shown in Fig 1.3.

The solutions are plotted between the two objectives in the figure, there may be more objectives. The pareto front is represented through solutions A - B - C - D - E - F - G. The pareto front is formed with solutions which are not dominated by other solutions; in other words, the solutions in the pareto front are better than their neighboring solutions on at least one front/axis. The solutions that are not included in the pareto front are those solutions which at least have one other solution in the population is better than itself in all the objectives. There are other solution methods involving more numbers of pareto fronts as well.

Multiobjective problems are more complex than the single objective problems. More calculations and analysis are required in implementing the multiobjective problem. Also, it should be understood, that the optimal solution is obtained based on the importance given to a certain objective and all objectives may not be optimized simultaneously.

1.5 OPTIMIZATION TECHNIQUES

The number of optimization techniques has been increasing with time. There is no specific count of the number of optimization techniques developed. The following chapters try to present a collection of optimization techniques which have been dominant in the research field with some new entrants as well.

Each chapter tries to capture the essence of the optimization technique in consideration. The chapters are organized in such a way that the interest in the technique

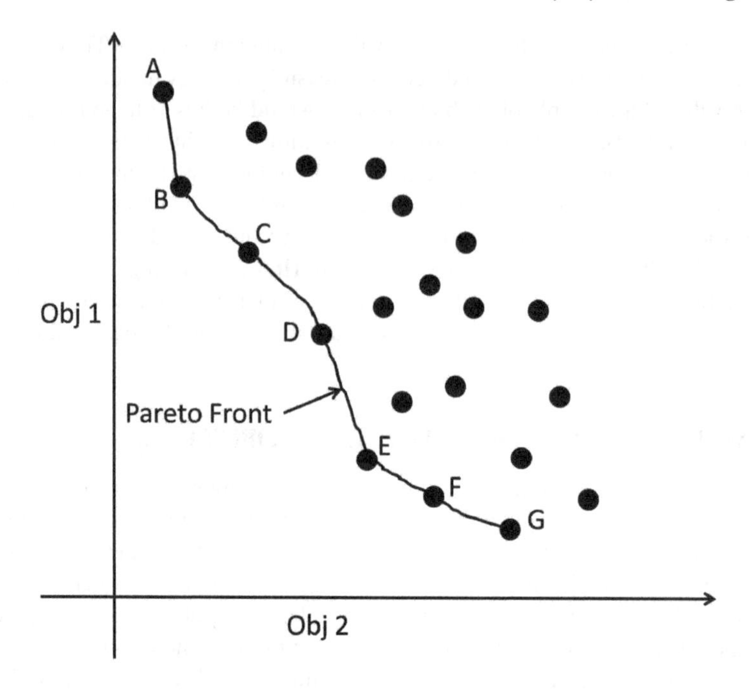

Figure 1.3: Multi-Objective Optimization

shall keep rising until the end of the chapter. Utmost care has been taken to make sure that the reader understands each and every step easily. The equations are solved with values put in them. The researchers or students working on these techniques can make this book a one-stop solution to all their optimization technique problems.

All the chapters are organized in a similar way to maintain uniformity. The chapters present a basic concept of the technique, along with their terminology and process, algorithm and flowchart, while finishing with an example solved for two iterations. The examples have a step-by-step solution for each optimization technique. In the last chapter, some of the techniques are collected for explanation. The second chapter is used to make the reader aware of the concepts of optimization problems. The techniques covered are:

Genetic Algorithm (GA): This technique is based on the evolution of species and takes inspiration from the fact that the individuals which are fittest will survive. The individuals which survive will be able to reproduce and therefore the next generation will have their chromosomes/traits.

Differential Evolution (DE): Differential Evolution implements processes similar to GA, however, the calculations are quite different. DE is applied to real values and not binary numbers with less number of calculations involved in it, thus making it fast.

Particle Swarm Optimization (PSO): Particle swarm optimization is derived from the movements of birds in a flock and fish in a school. The movement of these birds or fish is coordinated and aimed to optimize an individual's chances of finding prey or mates for reproduction and to avoid predators. It is also inspired from a human thought process which is a combination of personal experiences and social observation.
Artificial Bee Colony (ABC): Artificial Bee Colony derives its methodology from a hive of bees. This is a unique method within the optimization methods collection, as the bee does not represent the optimal solution; instead the bees are steps to improve the solutions. The method tries to add all parameters related to a bee hive, including the bee categories and their dancing. The bees are classified as employed bees, onlooker bees and scouts.
Shuffled Frog Leaping Algorithm (SFLA): As the name suggests, the method derives its workings from the frogs in a swamp. Shuffled Frog Leaping Algorithm is currently the trending optimization technique. It classifies the population into different memeplexes (groups) after sorting the frogs. The memeplexes are allowed to evolve on their own, then again collected back in a common population. The process is repeated again and again. The method has multiple calculations involved.
Grey Wolf Optimizer (GWO): The hunting methodology of wolves is applied to obtain the optimization method of Grey Wolf Optimizer. The hunt involves a number of leaders which lead the pack in the hunt. It is assumed that the leaders in the pack have the idea as to where the prey is and hence the pack follows these leaders.
Teaching Learning Based Optimization (TLBO): The process of Teaching Learning Based Optimization is derived from the process of teaching and learning in a class. The best solution is that the class/population acts as a teacher and helps other participants to improve their performance. A student in the class can also be inspired from the performance of other students and can improve their own performance with the help of these classmates. The method is further divided into different processes.
Bacteria Foraging Algorithm (BFA): Bacteria Foraging Algorithm is an algorithm based on the process of foraging, employed by the bacteria. Bacteria are small in size and multiply quickly. Their foraging methods involve attractants and repellants based on the environment, which may be inside a living organism
Whale Optimization: Whale optimization is inspired from the hunting strategy of the humpback whales. The humpback whales try and encircle their prey from all sides and block their paths of escape through a wall of bubbles. They move upward spirally while narrowing on their prey.
Bat Algorithm: Bat Algorithm is another optimization technique based on a hunting strategy, this time by the bat. Bats are of different types like fruit eating and hunting bats. The bats that hunt make use of the echolocation method to pinpoint the location of their prey. The bat sends out soundwaves which are incidental to their prey. The sound waves are then reflected back

toward the bat. The bat senses these reflected waves and analyzes them to pinpoint the location of their prey.

Firefly Algorithm: Firefly Algorithm copies the fireflies for optimization. Fireflies glow in the dark to attract mates and prey while repelling predators. There is a pattern in the glowing of fireflies. The brightness of the glow is considered to be the objective function value of that firefly.

Gravitational Search Algorithm (GSA): Gravitational Search Algorithm is based on Newton's laws. It involves gravity and other laws of motion. The movement of masses in this method is based on the force experienced by them.

Reducing Variable Trend Search (RVTS): Reducing Variable Trend Search works in a different way based on the methodologies involved. First, the best solutions are identified. Then their input variable values are varied to observe the trend in input variable values that give better results. These trends are used to reduce the search space. The processes are repeated to an extent until the search space is reduced to the size of a single solution, i.e., the optimal solution.

1.6 CONCLUSION

Optimization problems are a part of our day-to-day lives. For the simple ones, our smarts are enough to handle them but for the more complex ones, having large numbers of possible solutions, we need to utilize EOAs. This book deals with a number of such algorithms. The algorithms are based on different natural processes like evolution, hunting strategies, etc. Special focus has been kept on reader centric content arrangement. The solutions are provided in a step-by-step method for different algorithms. Also, on the other hand, a large number of references has been referred to through the process of writing the book.

The chapters are arranged in a similar manner for better understanding and adaptability by the users. All the chapters have a similar section, which first introduce the terminology related to the algorithm, then the process is expressed followed by pseudocode/algorithm and flowchart. The last section presents an example, which solves the algorithm in a step-by-step process for two iterations. Some of the chapters have some extra sections to handle the variants introduced for that method.

The next chapter is used to introduce different aspects required to understand the optimization algorithms.

2 Optimization Functions

2.1 INTRODUCTION

The optimization techniques need standard functions to be tested upon. In this chapter we will visit the standard optimization functions used to test the performance of various optimization techniques. The optimization functions are also solved for the ease of implementing them later. A large list of such optimization functions is also presented. As an addition, the graphs and ranges of input variables are also covered in the chapter. Lastly, the most commonly used optimization problems of the traveling salesman and hill climbing are discussed. The aim of this chapter is to set up the stage for understanding and implementing various optimization techniques covered in the later chapters.

2.2 STANDARD OPTIMIZATION FUNCTIONS

A number of EOAs have been introduced in the research domain from time to time. These are derived from some natural process and have their applications in varied fields of engineering. For any given case, the EOA may be fine-tuned to serve the specific problem of optimization, but not other optimization problems. The performance of these EOAs can therefore be best compared against a common problem. This section presents a list of some of the standard optimization problems. All the optimization problems cannot be enlisted here, however, a large number of standard optimization problems are listed in [6]. Also the range of input variables may vary based on different resources. These functions are, in general, continuous, convex and can be applied for 'n' dimensions.

Sphere Function

The sphere function is represented as:

$$f(x) = \sum_{i=1}^{n} x_i^2 \tag{2.1}$$

where $x_i \in [-5.12, 5.12]$. The complexity of sphere function is low and any optimization technique should be able to find the optimal value quite accurately. The sphere function for two dimensions can be represented as shown in Fig 2.1.

Ackley Function

$$f(x) = 20 + e - 20exp\left(-0.2\sum_{i=1}^{n}\frac{x_i^2}{n}\right) - exp\left(\sum_{i=1}^{n}\frac{(cos2\pi x_i)}{n}\right) \tag{2.2}$$

where $x_i \in [-32, 32]$. Fig 2.2 shows Ackley function for two dimension with a range of -32 to 32.

DOI: 10.1201/b22647-2

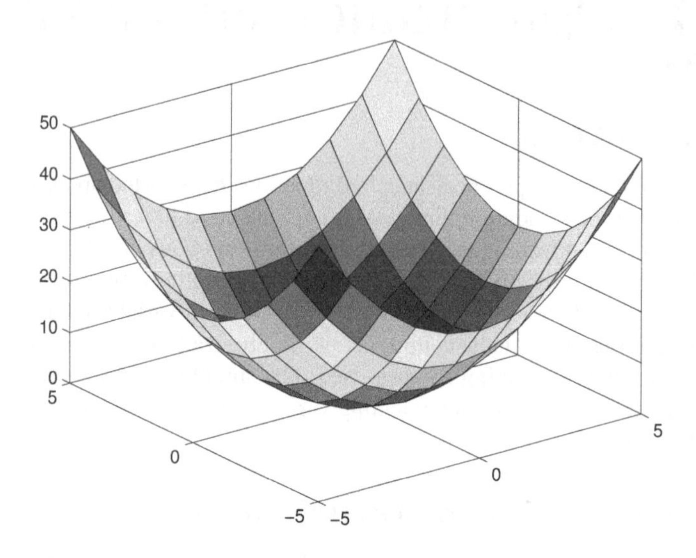

Figure 2.1: Sphere Function

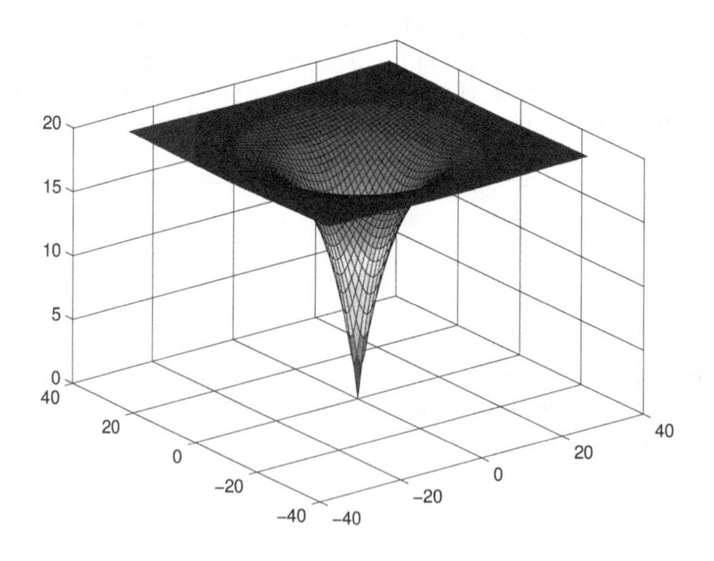

Figure 2.2: Ackley Function

Rosenbrock Function

$$f(x) = \sum_{i=1}^{n-1} [100(x_{i+1} - x_i^2)^2 + (x_i - 1)^2] \tag{2.3}$$

where $x_i \in [-10, 10]$. Fig 2.3 shows the Rosenbrock function for two dimension with a range of -10 to 10.

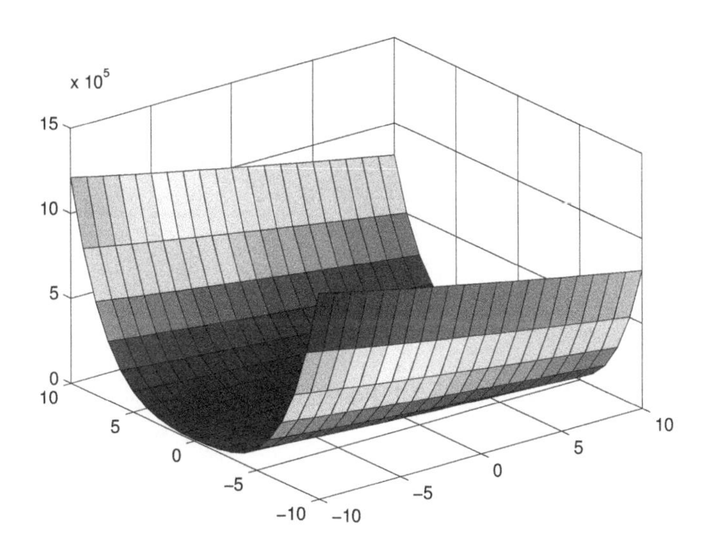

Figure 2.3: Rosenbrock Function

Rastrigin Function

$$f(x) = 10n + \sum_{i=1}^{n} (x_i^2 - 10cos(2\pi x_i)) \tag{2.4}$$

where $x_i \in [-5.12, 5.12]$. Fig 2.4 shows the Rastrigin function for two dimension with a range of -5 to 5.

Griewank Function

$$f(x) = 1 + \sum_{i=1}^{n} \left(\frac{x_i^2}{4000} \right) - \Pi_{i=1}^{n} cos \left(\frac{x_i}{\sqrt{i}} \right) \tag{2.5}$$

where $x_i \in [-600, 600]$. Fig 2.5 shows the Griewank function in two dimension with a range of -5 to 5.

Fletcher Powell Function

$$f(x) = \sum_{i=1}^{n} (A_i - B_i)^2 \tag{2.6}$$

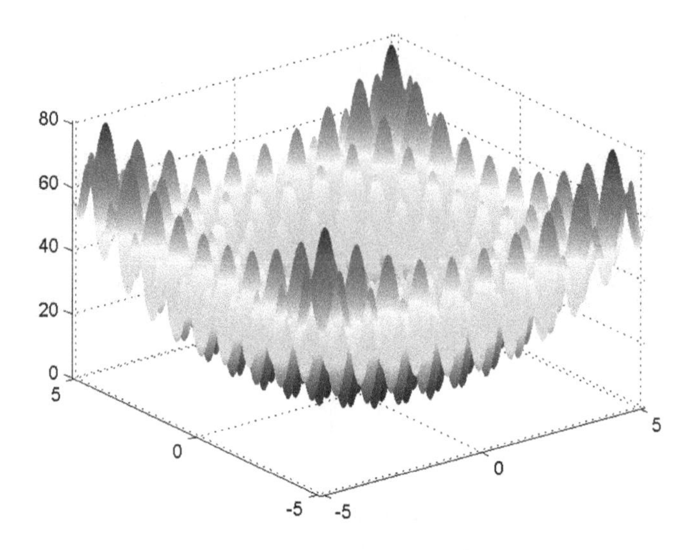

Figure 2.4: Rastrigin Function

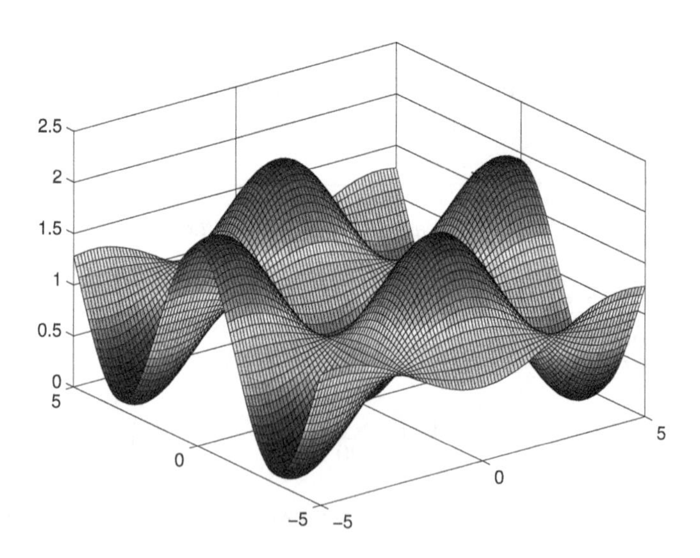

Figure 2.5: Griewank Function

where

$$A_i = \sum_{i=1}^{n} (a_{ij} sin\alpha_j + b_{ij} cos\alpha_j)$$

$$B_i = \sum_{i=1}^{n} (a_{ij} sinx_j + b_{ij} cosx_j)$$

where $\alpha_i \in [-\pi, \pi]$ such that $i \in [1, 2, ...n]$ and
$a_{ij}, b_{ij} \in [-100, 100]$ such that $i, j \in [1, 2, ...n]$

Zakharov Function

$$f(x) = \sum_{i=1}^{n} x_i^2 + (\sum_{i=1}^{n} 0.5ix_i)^2 + (\sum_{i=1}^{n} 0.5ix_i)^4 \tag{2.8}$$

where $x_i \in [-5, 10]$. Fig 2.6 shows a two dimensional representation of Zakharov function with a range of −5 to 5.

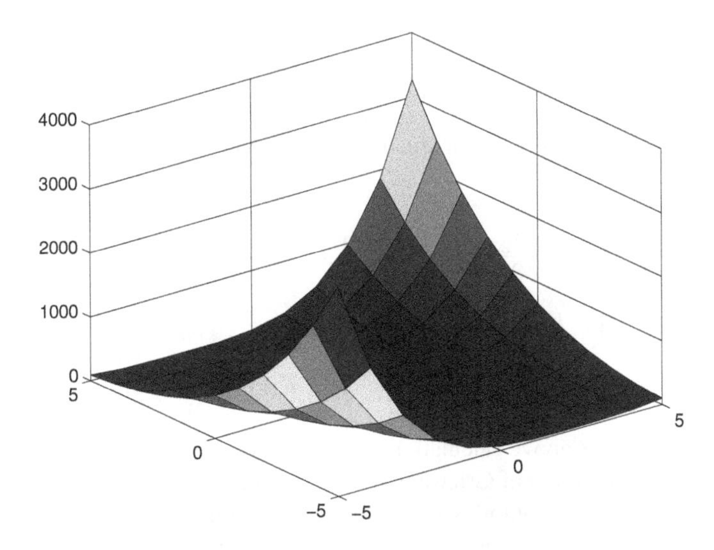

Figure 2.6: Zakharov Function

The above is just a sample list of functions. Now let us see how the values are calculated for Sphere and Griewank functions. The reader can then extend it to other functions in a similar way. Evaluation of optimization problems function value shall help us understand the solutions provided for EOA in the later chapters.

Let us assume that the random values of the input variables are: [−1, 4].

The input variable are 2, thus n = 2. The number of input variables can be increased and so will the value of 'n'. For Sphere function, the objective function value will be

$$f(x) = \Sigma_{i=1}^{n} x_i^2 = x_1^2 + x_2^2 = (-1)^2 + 4^2 = 17.$$

Similarly for the range of −5 to 5, the different values for the objective function of Sphere problem would be as given in Table 2.1:

Table 2.1
Objective Function Values for Sphere Function

	−5	**−4**	**−3**	**−2**	**−1**	**0**	**1**	**2**	**3**	**4**	**5**
−5	50	41	34	29	26	25	26	29	34	41	50
−4	41	32	25	20	17	16	17	20	25	32	41
−3	34	25	18	13	10	9	10	13	18	25	34
−2	29	20	13	8	5	4	5	8	13	20	29
−1	26	17	10	5	2	1	2	5	10	17	26
0	25	16	9	4	1	0	1	4	9	16	25
1	26	17	10	5	2	1	2	5	10	17	26
2	29	20	13	8	5	4	5	8	13	20	29
3	34	25	18	13	10	9	10	13	18	25	34
4	41	32	25	20	17	16	17	20	25	32	41
5	50	41	34	29	26	25	26	29	34	41	50

Now for a Griewank problem, the objective function value for an input of [−1,4] can be found as:

$$f(x) = 1 + \left(\frac{x_1^2}{4000} + \frac{x_2^2}{4000}\right) - \left(cos\left(\frac{x_1}{\sqrt{1}}\right) * cos\left(\frac{x_2}{\sqrt{2}}\right)\right)$$
$$= 1 + \left(\frac{-1^2}{4000} + \frac{4^2}{4000}\right) - \left(cos\left(\frac{-1}{\sqrt{1}}\right) * cos\left(\frac{4}{\sqrt{2}}\right)\right)$$
$$= 1 + 0.0042 - (-0.514) = 1.5182 \quad (2.9)$$

According to the above calculation, for the range of −5 to 5, the different values for the objective function of Griewank problem would be as given in Table 2.2:

Table 4 (Appendix) compiles different standard optimization functions while providing their names, number of dimensions, range and the equation.

2.3 TRAVELING SALESMAN PROBLEM

A Traveling Salesman Problem (TSP) is an optimization problem consisting of a salesman, cities and the distances amongst these cities. The salesman has to visit all the cities exactly once, but tries to minimize the distance travelled. Sometimes, the distance is replaced by cost and then the objective of the problem is to minimize the cost instead of the distance. In other cases, the same problem is converted to minimize the time required to travel all the cities. In any case, the salesman has to travel to all the cities to reduce time or cost or distance or a combination of these

Table 2.2
Objective Function Values for Griewank Function

	−5	−4	−3	−2	−1	0	1	2	3	4	5
−5	1.2744	1.2801	1.1568	0.9630	0.7908	0.7225	0.7908	0.9630	1.1568	1.2801	1.2744
−4	0.4066	0.3861	0.6643	1.1069	1.5011	1.6576	1.5011	1.1069	0.6643	0.3861	0.4066
−3	0.0943	0.0644	0.4866	1.1576	1.7551	1.9922	1.7551	1.1576	0.486	0.0644	0.0943
−2	0.6229	0.6090	0.7855	1.0668	1.3176	1.4171	1.3176	1.0668	0.7855	0.6090	0.6229
−1	1.5054	1.5182	1.285	0.9169	0.5897	0.4599	0.5897	0.9169	1.2851	1.5182	1.5054
0	1.9296	1.9553	1.5253	0.8450	0.2400	0	0.2400	0.8450	1.5253	1.9553	1.9296
1	1.5054	1.5182	1.2851	0.9169	0.5897	0.4599	0.5897	0.9169	1.2851	1.5182	1.5054
2	0.6229	0.6090	0.7855	1.0668	1.3176	1.4171	1.3176	1.0668	0.7855	0.6090	0.6229
3	0.0943	0.0644	0.4866	1.1576	1.7551	1.9922	1.7551	1.1576	0.4866	0.0644	0.0943
4	0.4066	0.3861	0.6643	1.1069	1.5011	1.6576	1.5011	1.1069	0.6643	0.3861	0.4066
5	1.2744	1.2801	1.1568	0.9630	0.7908	0.7225	0.7908	0.9630	1.1568	1.2801	1.2744

three factors. The TSP has also been modified to solve the problem of goods delivery and pickup. The TSP is solved using greedy algorithms and optimization techniques.

The TSP problem was initially introduced by W. R. Hamilton, an Irish mathematician, and Thomas Kirkman, a British mathematician, in the 1800s. Karl Menger has contributed greatly to the evolution of TSP. The name TSP was given by Hassler Whitney at Princeton University. TSP is a peculiar case of an "NP hard" optimization problem. A detailed survey of different versions of TSP problems and their timelines are presented in [74] and [137].

If the salesman has to visit N cities there would be (N-1)! possibilities. For example, there are four cities, A, B, C and D. Thus, the different possible routes would be: [A - B - C - D - A], [A - B - D - C - A], [A - C - B - D - A], [A - C - D - B - A], [A - D - C - B - A] and [A - D - B - C - A], considering A as the hometown. In the various combinations presented, it can be observed that the routes can be paired in twos (which are exactly opposite in combination like ABCDA and ADCBA) and their results will be similar; this will be shown again in Table 2.3.

The problem of TSP has been expanded for thousands of cities and their solutions have been formulated. In TSP it is compulsory that the salesman should have a hometown from where he starts and returns after every trip. The best part of TSP is that no mathematical background is required for a researcher to understand and implement the problem.

Let us take a simple example with four cities as given in Fig 2.7

The hometown for the salesman is A. The numbers on the edges give the time or distance or cost (or a combination of all) for traveling through that edge. Thus, now we will examine all the possible routes. The total weightage of a route shall be the sum of all the weights of edges it travels. The objective function value for each path is presented in Table 2.3.

From Table 2.3, it can be found that routes [A - B - C - D - A] or [A - D - C - B - A] are optimal routes with an objective function value of 80. But these routes are

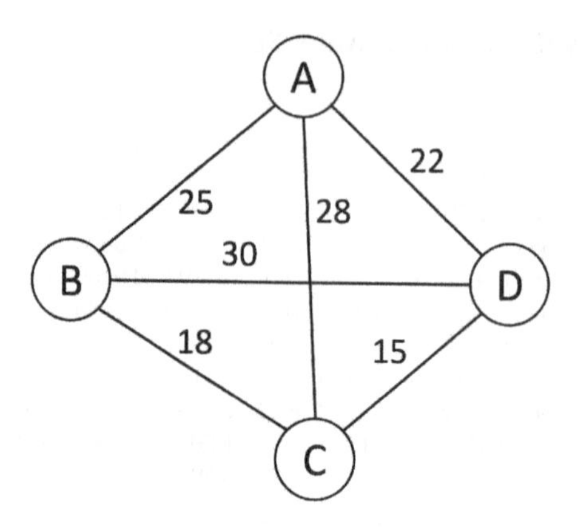

Figure 2.7: Simple Traveling Salesman Problem

exactly opposite to each other. Since the edges travelled are the same, the objective function remains the same. This will also be true for routes [A - B - D - C - A] and [A - C - D - B - A] and the remaining pair of routes. It can be concluded that for this simple version of TSP the number of distinct possiblities are actually (N-1)!/2. However, this may not be the case for all different versions of TSP, as the cost of traveling in a particular direction may be different than the cost of traveling in the opposite direction.

The mathematical model for a general TSP can be represented as in [52] (there are many variants of TSP and each one has its own objective function and

Table 2.3
Objective Function Values for Traveling Salesman Problem

Route	Objective Function Value
A - B - C - D - A	25 + 18 + 15 + 22 = 80
A - B - D - C - A	25 + 30 + 15 + 28 = 98
A - C - B - D - A	28 + 18 + 30 + 22 = 98
A - C - D - B - A	28 + 15 + 30 + 25 = 98
A - D - B - C - A	22 + 30 + 18 + 28 = 98
A - D - C - B - A	22 + 15 + 18 + 25 = 80

constraints):

$$ObjectiveFunction = min \sum c_{ij}x_{ij} \tag{2.10a}$$

$$s.t. \sum_{j=1}^{n} x_{ij} = 1 \qquad i = 1,2,....n \tag{2.10b}$$

$$\sum_{i=1}^{n} x_{ij} = 1 \qquad j = 1,2,....n \tag{2.10c}$$

$$x_{ij} \in 0,1 \qquad i,j = 1,2,....n \quad i \neq j \tag{2.10d}$$

In the above equations, Equation 2.10a gives the objective function for a TSP. In Equation 2.10a, c_{ij} represents the cost or distance or time (or combination of all) required to travel from city 'i' to city 'j' and x_{ij} is a decision variable, which can have a value of 0 or 1 (2.10d). $x_{ij} = 0$ means that the route between city 'i' and city 'j' was not taken and for $x_{ij} = 1$, the salesman traveled from city 'i' to city 'j'. Equations 2.10b and 2.10c, are contraints to make sure that the salesman enters a city once and leaves it once, respectively. It also means that all the cities shall be visited exactly once.

Some of the variants and applications of TSP [52], [27], [74] are presented below:

Variants

- sTSP: symmetric salesman traveling problem
- aTSP: asymmetric salesman traveling problem
- mTSP: multisalesman traveling problem
- bTSP: bottleneck traveling salesman problem
- cTSP: clustered traveling salesman problem
- pTSP: period traveling salesman problem
- tdTSP: time dependent traveling salesman problem

Applications

- VRP: Vehicle Routing Problem
- PCB Drilling
- X-ray Crystallography
- Machine Scheduling and Sequencing
- Computer Wiring
- Genome Sequencing [7]

A survey of TSP Problems is presented in [3]. A collection of research papers which have presented a solution for various number of cities for the TSP problem is suggested. A collection of algorithms used to solve the TSP problem is also presented. These algorithms are:

- Brute Force Method
- Greedy Approach
- Nearest Neighbor Heuristic
- Branch and Bound Method

- 2-Opt Algorithm
- Greedy 2-Opt Algorithm
- Genetic Algorithm
- Simulated Annealing (SA)
- Neural Network

Some really useful information is available at [1] along with the required database related to various problems of TSP.

2.4 HILL CLIMBING

Hill climbing is a simple optimization problem. In fact, it is suggested to the readers to first go through this problem before studying optimization techniques. Hill climbing relates to finding the topmost or the most heighted position in a given search area. The solution of the problem is to find the peak of the tallest hill and, therefore, it will usually be a maximization problem [18]. In some cases/algorithms, it is assumed that the position of the peak is known. This section is added in the book to analyze the simple issues arising in an optimization problem.

Figure 2.8 shows a simple hill-climbing surface and is the realization of the equation $e^{-(x^2+y^2)}$, where the value of x and y vary from -2 to 2.

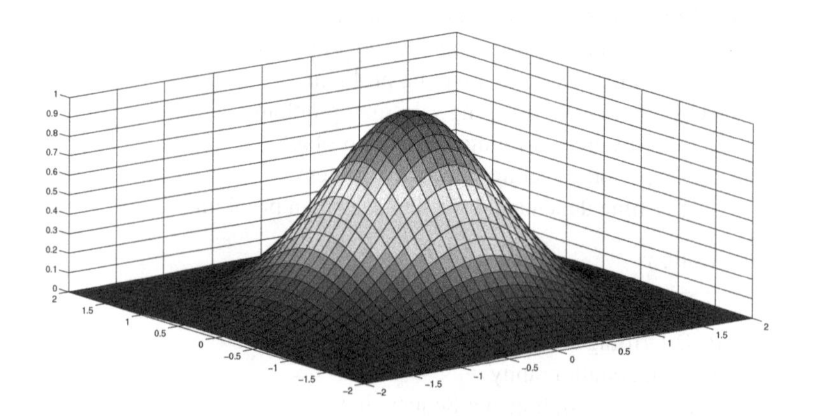

Figure 2.8: Simple Hill Climbing Problem

Another version of the Hill Climbing problem is given in Fig 2.9. It can be observed that there is a smaller peak in the neighborhood of a larger peak. The local optima problem starts with an optimization problem getting stuck with the smaller peak and assuming it to be the optimal solution. The inability of the optimization technique to reach the optimal solution is one of the most important issues to be handled by an EOA. Existence of such multiple local optima/small hills makes the problem more and more complex. The equation for generating the surface shown in Fig 2.9 is given by $e^{-(x^2+y^2)}+2e^{-((x-1.7)^2+(y-1.7)^2)}$ in the range of -2 to 4.

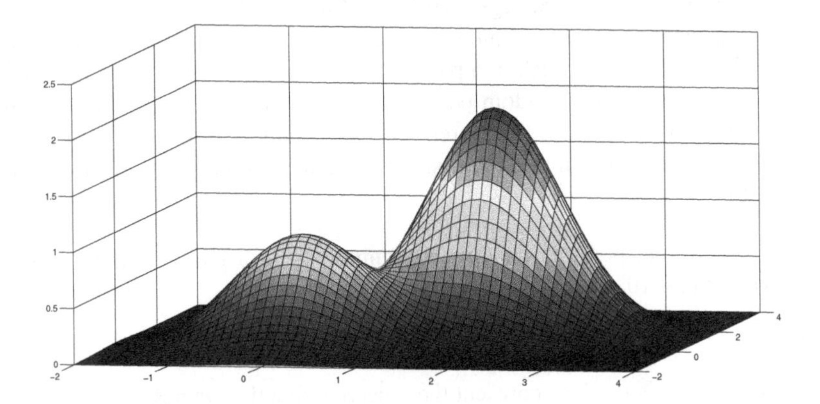

Figure 2.9: Multiple Hills in an Hill Climbing Problem

Hill Climbing can be solved through different optimization techniques like Simulated Annealing, Genetic Algorithm, etc. Let us, however, discuss some methods dedicated to solving the Hill Climbing problem itself. From the different Hill Climbing techniques, it is assumed in some that the best solution is known. The different Hill Climbing solution techniques are:

1. **Simple Hill Climbing**: In a simple hill climbing algorithm, neighboring nodes are examined one after another. The first node, which gives a better objective function value than the existing position, is selected. The same process shall be repeated for the next position and so on. Simple Hill Climbing can be best applied to a system as represented by Fig 2.8.
2. **Steepest Ascent Hill Climbing**: In the steepest ascent hill climbing technique all the neighboring positions are evaluated. The position which gives the largest improvement in the objective function value is selected as the next node.
3. **Iterative Hill Climbing [18]**: In this technique, the process starts with a simple random guess. Increments are done for one input variable of the random solution. If the objective function value improves we continue to move in the same direction, by increasing the value of the input variable. In case the objective function value reduces, we undo the previous step and start increasing the value of another input variable. The process continues until all the variables have been tested. The value of the increment is then halved and the sign is reversed, while repeating the whole process again. The process repeats until the increments have been halved for considerable number of times.
4. **Stochastic Hill Climbing**: This method chooses a neighboring position randomly and evaluates its objective function value. The probability of this random position being selected as the next position depends on the steepness of the move.

5. **Random Restart/Shotgun Hill Climbing**: In the random restart hill climbing method, the hill climbing is started with a random initial position. Once the method gets stuck at a particular position, it again restarts the hill climbing from another random position. This method can be combined with other methods as well for the search process. It may also be noted that it may find the optimal position in the first run also, and can be repeated for a number of times.

The different possible conditions/terminology in a Hill Climbing problem are shown in Fig 2.10. Let us list the terminology involved:

X-Axis Denotes the various values that the input variable can have.

Y-Axis Y-axis is used to represent the objective function values.

Local Optima/Maxima It is a position whose neighboring positions have a lower objective function value, however, this position does not have the best objective function value in the whole search space.

Global Maximum It is a position whose objective function value is the highest/-maximum/optimal and no other position in the search space has a better objective function value.

Plateau It is a region in the search space where the objective function value does not change by much and the positions in the neighboring of this region have lower objective function values.

Shoulder It is a plateau on a slope of a hill.

Ridge/Alley Ridge is a long narrow hilltop in the search region. Alley is the reverse of ridge and represents a long narrow valley.

Let us try and solve the sphere function through some of the above mentioned Hill Climbing Techniques. However, we have explained the sphere function as a minimization function in Eq 2.1. Since Hill Climbing is used for maximization problems, we will convert it into a maximization problem by taking the values as $\frac{1}{1+f(x)}$. The sphere function with the above change in its objective function can now be shown as given in Fig 2.11.

The values of the Sphere function with the above mentioned changes for maximization is given in Table 2.4.

Now to apply some of the Hill Climbing Techniques to the Inverse Sphere function described above.

1. *Simple Hill Climbing:* The initial solution is assumed to be [4,−4], whose objective function value is 0.0303. In all there would be 8 neighboring states to the current state (The number of states would vary depending on the number of input variables). These neighboring states can be reached by adding or subtracting one step value from the input variable values of the current state. First

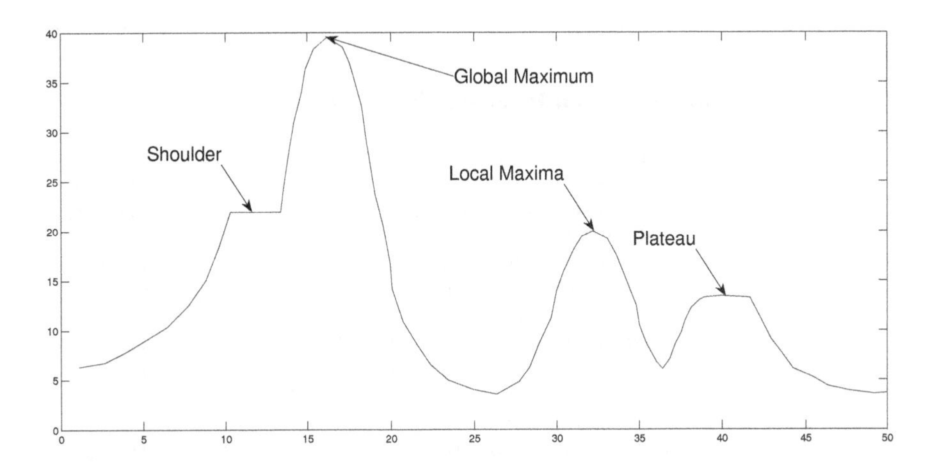

Figure 2.10: Terminology of the Hill Climbing Problem

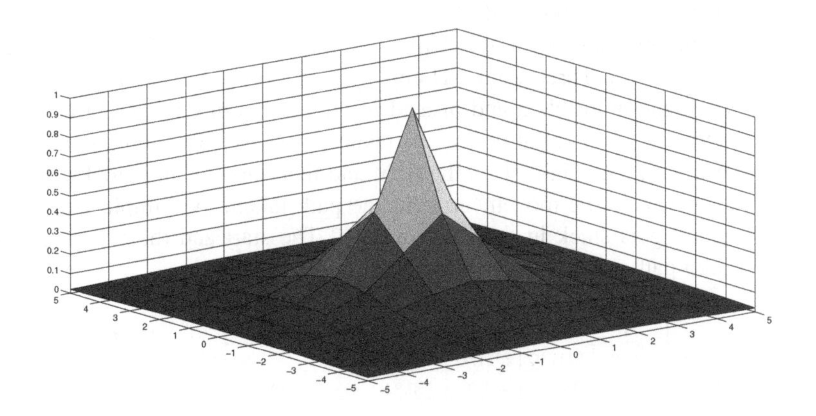

Figure 2.11: Maximization of Sphere Function

neighboring state being considered is [5,−4], whose objective function value is 0.0238. Since it is less, we move to the next neighboring state as [5,−3], the objective function value is 0.0286 which is still less than 0.0303 of the current state. Next neighboring state is [4,−3], having an objective function value of 0.0385. Since this objective function value is greater than the objective function value of the current state, the current state is now moved to [4,−3]. A neighboring state to [4,−3] will then be found having a better objective function value and the current state shall be moved to first neighboring state evaluated as having a better objective function value. The process will

Table 2.4
Objective Function Values for Inverse Sphere Function

	−5	−4	−3	−2	−1	0	1	2	3	4	5
−5	0.02	0.0238	0.0286	0.0333	0.037	0.0385	0.037	0.0333	0.0286	0.0238	0.02
−4	0.0238	0.0303	0.0385	0.0476	0.0556	0.0588	0.0556	0.0476	0.0385	0.0303	0.0238
−3	0.0286	0.0385	0.0526	0.0714	0.0909	0.1	0.0909	0.0714	0.0526	0.0385	0.0286
−2	0.0333	0.0476	0.0714	0.1111	0.1667	0.2	0.1667	0.1111	0.0714	0.0476	0.0333
−1	0.037	0.0556	0.0909	0.1667	0.3333	0.5	0.3333	0.1667	0.0909	0.0556	0.037
0	0.0385	0.0588	0.1	0.2	0.5	1	0.5	0.2	0.1	0.0588	0.0385
1	0.037	0.0556	0.0909	0.1667	0.3333	0.5	0.3333	0.1667	0.0909	0.0556	0.037
2	0.0333	0.0476	0.0714	0.1111	0.1667	0.2	0.1667	0.1111	0.0714	0.0476	0.0333
3	0.0286	0.0385	0.0526	0.0714	0.0909	0.1	0.0909	0.0714	0.0526	0.0385	0.0286
4	0.0238	0.0303	0.0385	0.0476	0.0556	0.0588	0.0556	0.0476	0.0385	0.0303	0.0238
5	0.02	0.0238	0.0286	0.0333	0.037	0.0385	0.037	0.0333	0.0286	0.0238	0.02

similarly be repeated until the optimal state is reached and there are not better states in its neighborhood.

Another example of Simple Hill Climbing is shown in Fig 2.12. The dataset represents the Three Camel Hump optimization problem. However, this is a minimization problem and instead of climbing a hill, a descent in the valley is required. The initial position is taken as [4, −4]. The neighborhood search starts with the neighbor on the right and then moving anti clockwise. As can be seen from the movement from the path taken, the Simple Hill Climbing method gets stuck in the local minima. The start and the end points of the search path are denoted by bold.

	-5	-4	-3	-2	-1	0	1	2	3	4	5
-5	2047.92	2033.92	2021.92	2011.92	2003.92	1997.92	1993.92	1991.92	1991.92	1993.92	1997.92
-4	490.87	477.87	466.87	457.87	450.87	445.87	442.87	441.87	442.87	445.87	450.87
-3	94.45	82.45	72.45	64.45	58.45	54.45	52.45	52.45	54.45	58.45	64.45
-2	36.87	25.87	16.87	9.87	4.87	1.87	0.87	1.87	4.87	9.87	16.87
-1	31.12	21.12	13.12	7.12	3.12	1.12	1.12	3.12	7.12	13.12	21.12
0	25.00	16.00	9.00	4.00	1.00	0.00	1.00	4.00	9.00	16.00	25.00
1	21.12	13.12	7.12	3.12	1.12	1.12	3.12	7.12	13.12	21.12	31.12
2	16.87	9.87	4.87	1.87	**0.87** ←	1.87	4.87	9.87	16.87	25.87	36.87
3	64.45	58.45	54.45	52.45	↗ 52.45	54.45	58.45	64.45	72.45	82.45	94.45
4	450.87	**445.87** →	442.87 →	441.87	442.87	445.87	450.87	457.87	466.87	477.87	490.87
5	1997.92	1993.92	1991.92	1991.92	1993.92	1997.92	2003.92	2011.92	2021.92	2033.92	2047.92

Figure 2.12: Simple Hill Climbing

2. *Steepest Ascent Hill Climbing:* The initial state is again taken as [4, −4] with an objective function value of 0.0303. It has eight neighboring states as can be observed in Table 2.4. These states are [3,−3], [3,−4], [3,−5], [4,−3], [4,−5],

[5,−3], [5,−4] and [5,−5]. The objective function values of these states can be checked in Table 2.4. The steepest rise in the objective function value comes if the next state is chosen as [3,−3] with an objective function value of 0.0526. This step shall be repeated again and again until the optimal solution is reached. It can be observed that this method is quite faster than Simple Hill Climbing Technique.
Similar to the Simple Hill Climbing method, the Steepest Ascent Hill Climbing method is denoted by Fig 2.13 for the Three Camel Hump optimization problem. As can be seen, Steepest Ascent Hill Climbing method takes a shorter route, but gets stuck in the same local minima as the Simple Hill Climbing method.

	-5	-4	-3	-2	-1	0	1	2	3	4	5
-5	2047.92	2033.92	2021.92	2011.92	2003.92	1997.92	1993.92	1991.92	1991.92	1993.92	1997.92
-4	490.87	477.87	466.87	457.87	450.87	445.87	442.87	441.87	442.87	445.87	450.87
-3	94.45	82.45	72.45	64.45	58.45	54.45	52.45	52.45	54.45	58.45	64.45
-2	36.87	25.87	16.87	9.87	4.87	1.87	0.87	1.87	4.87	9.87	16.87
-1	31.12	21.12	13.12	7.12	3.12	1.12	1.12	3.12	7.12	13.12	21.12
0	25.00	16.00	9.00	4.00	1.00	0.00	1.00	4.00	9.00	16.00	25.00
1	21.12	13.12	7.12	3.12	1.12	1.12	3.12	7.12	13.12	21.12	31.12
2	16.87	9.87	4.87	↗ 1.87	→ **0.87**	1.87	4.87	9.87	16.87	25.87	36.87
3	64.45	58.45	↗ 54.45	52.45	52.45	54.45	58.45	64.45	72.45	82.45	94.45
4	450.87	**445.87**	442.87	441.87	442.87	445.87	450.87	457.87	466.87	477.87	490.87
5	1997.92	1993.92	1991.92	1991.92	1993.92	1997.92	2003.92	2011.92	2021.92	2033.92	2047.92

Figure 2.13: Steepest Ascent Hill Climbing

3. *Iterative Hill Climbing:* Assume the initial position as [4,−4] again. The step size is assumed to be 2. Let us increase the first variable by 2, which is not possible due to the constraints, so we increase it to the maximum limit, i.e., 5. The new state to be evaluated would be [5,−4] having objective function value of 0.0238 which is less than [4,−4]. So we retain the current state as [4,−4]. Next we start increasing the second variable. Thus now, the next state to be considered should be [4,−2], which has an objective function value of 0.0476. Since the objective function value has increased from 0.0303 to 0.0476, the second variable should again be increased by 2. The next state would be [4,0] having an objective function value of 0.0588, which is again an improvement from the previous position. The current state is now moved to [4,0]. In the next step we move to [4,2], whose objective function value is 0.0476. Since this is not an improvement, we return back to [4,0] as the current state. Now we reduce the first variable by 1 to move to [3,0] with an objective function value of 0.1. As this is an improvement, we move further along the same path and move to [2,0] with an objective function value of 0.2. Again reduce the first variable to move to [1,0], giving an objective function value of 0.5 and one more step to [0,0] having a value of 1. We need to further reduce the first variable by 1 to move to [−1,0] but now the objective function value will fall to

0.5. Thus this step is cancelled and we retain the current state at [0,0]. We now need to reduce the second variable by 1 and evaluate [0,−1] which is having 0.5 as its objective function value. We cannot move to the new state. As the steps cannot be further reduced we conclude the optimal state to be [0,0]. It is advised to go through the path of the current state in Table 2.4.
In the Iterative Hill Climbing method, applied to the Three Camel Hump optimization problem, we start with an increase in first variable by 2, then second variable by 2, then decrease both of them in the same order. The path taken by Iterative Hill Climbing method is shown in Fig 2.14. In this method as well, the end position is a local optima.

	-5	-4	-3	-2	-1	0	1	2	3	4	5
-5	2047.92	2033.92	2021.92	2011.92	2003.92	1997.92	1993.92	1991.92	1991.92	1993.92	1997.92
-4	490.87	477.87	466.87	457.87	450.87	445.87	442.87	441.87	442.87	445.87	450.87
-3	94.45	82.45	72.45	64.45	58.45	54.45	52.45	52.45	54.45	58.45	64.45
-2	36.87	25.87	16.87	9.87	4.87	1.87	0.87	1.87	4.87	9.87	16.87
-1	31.12	21.12	13.12	7.12	3.12	1.12	1.12	3.12	7.12	13.12	21.12
0	25.00	16.00	9.00	4.00	1.00	0.00	1.00	4.00	9.00	16.00	25.00
1	21.12	13.12	7.12	3.12	1.12	1.12	3.12	7.12	13.12	21.12	31.12
2	16.87	9.87	4.87	1.87	**0.87**	1.87	4.87	9.87	16.87	25.87	36.87
3	64.45	58.45	54.45	52.45	52.45	54.45	58.45	64.45	72.45	82.45	94.45
4	450.87	**445.87**	442.87	441.87	442.87	445.87	450.87	457.87	466.87	477.87	490.87
5	1997.92	1993.92	1991.92	1991.92	1993.92	1997.92	2003.92	2011.92	2021.92	2033.92	2047.92

Figure 2.14: Iterative Hill Climbing

As a last part of Hill Climbing Methods, let us examine how the Algorithms for the above discussed methods can be developed.

1. *Simple Hill Climbing:* In Simple Hill Climbing method, the initial position in the search space is first determined. Assuming that the problem is a two-dimensional problem, a nested loop is utilized. The nested loop should cover all the neighboring positions. In the loops, '1' is considered the step size. Also the outer loop variable is 'i' whereas 'j' is the inner loop variable. Inside the nested loop we evaluate the objective function value of the neighboring position given by [x+i] and [y+j]. If the objective function value of the new position is better than that of the existing position, then the search process shifts to the new position and the search continues. In case, none of the neighboring positions have a better objective function value then, the search process ends. Algorithm 2.1 gives the algorithm for Simple Hill Climbing method.

Algorithm 2.1 Simple Hill Climbing

```
initialize position [x y]
label A
for -1 to 1 do
   for -1 to 1 do
      Evaluate objective function value at [x+i y+j]
      if obj value ([x+i y+j]) > obj value ([x y]) then
         x=x+i
         y=y+j
         go to A
      end if
   end for
end for
```

2. *Steepest Ascent Hill Climbing:* Algorithm 2.2 presents the algorithm for Steepest Ascent Hill Climbing. This method is similar to Simple Hill Climbing method until the nesting of loops. In the loops, the step size is assumed as 1 and the outer and inner loop variables are 'i' and 'j', respectively. However, in Steepest Ascent Hill Climbing method, all the neighboring positions are evaluated. The position of the search will shift to the position which provides the best objective function value of all the neighboring positions. In other words, the best objective function value of all the neighboring positions is compared with the current position, if it is better than the current position, then the search process will continue from this new position.

Algorithm 2.2 Steepest Ascent Hill Climbing

```
initialize position [x y]
label A
for -1 to 1 do
   for -1 to 1 do
      evaluate and store Objective function value at [x+i y+j]
   end for
end for
[x‘ y‘] = max (Objective Function value [x+i y+j])
if obj value ([x‘ y‘]) > obj value ([x y]) then
   x=x‘
   y=y‘
   go to A
end if
```

3. *Iterative Hill Climbing:* Iterative Hill Climbing method is a little complicated as compared to the above two methods. Its algorithm is given in Algorithm 2.3. In Iterative Hill Climbing method, we need not only initialize the initial position but also the initial step size and the minimum step size (below which

the search process will stop). The initial loop makes sure that the current step size is greater than or equal to the specified minimum step size. The next inner loop is utilized to decide the search direction of the process, assuming that this is a two-dimensional search. Also this loop variable is taken as 'i'. In this loop one of the variable values is varied and then we evaluate the objective function value at the new position. In case the objective function value of the new position is better than the current position, then the search shifts to the new position and the direction is retained (i=i-1). On the other hand, if the objective function value of the new position is not better than the current position, than the direction shall change in the next iteration. Also, if the search process has exhausted all the search directions then the step size is reduced and the search process continues. The search process will continue until the step size reduces to below the specified minimum step size.

Algorithm 2.3 Iterative Hill Climbing

```
initialize position [x y], step size, step size min
for step size>=step size min do
  for 1 to 4 do
    if i==1 then
      x'=x + step size
    else if i==2 then
      y'=y + step size
    else if i==3 then
      x'=x - step size
    else
      y'=y - step size
    end if
    evaluate Objective function value at [x' y']
    if obj value ([x' y']) > obj value ([x y]) then
      x=x'
      y=y'
      i=i-1
    end if
  end for
  step size = step size/2
end for
```

The remaining methods cannot be solved in such an easy way and random numbers would have to be dealt with. Thus those methods are not numerically solved here.

3 Genetic Algorithm

3.1 INTRODUCTION

The EOA techniques are based on various biological processes observed in nature. The best process of optimal behavior observed in nature is the evolution of species through their generations. Each new generation of any species will keep changing from the previous one and will try to best adapt to the changing environment. The features appearing in the new generation will make it more adept to survive.

Darwinian theory proposed by Charles Darwin in 1859 mentions, "Survival of the fittest", which means that the individuals which can 'best adapt' to the environment shall survive. These individuals are best suited to live in this world and are able to reproduce to create a new generation. Thus only the individuals which shall survivie shall reproduce and pass on their genes/traits to the next generations. Also in a reverse observation, the individuals who are unable to utilize the resources optimally will have lesser and lesser descendants and their genes/traits shall diminish over a period of time.

In nature, we find different kinds of species which have developed over time for peculiar tasks like the cheetah, which is the fastest running animal on earth. Over a period of centuries, cheetahs, which were able to run faster, survived and reproduced and their offspring ran fast. Further, the offspring of cheetahs that ran fast survived and reproduced. This process repeated over a period of multiple generations, producing cheetahs as the fastest animals on earth. It also developed other characteristics like the fast swallowing of meat, managing high heart rate, etc. which were necessary for it to survive. Here, survival was the objective and the objective function was to run fast. The faster a cheetah could run, the fitter it became and the more its chances of passing on the genes to the next generation. This same methodology has been implemented in the Genetic Algorithm.

Genetic Algorithm (GA) tries to implement the same evolving behavior/process of nature to find the optimal solution. GA was one of the first bioinspired EOA. It can also be observed from overall research publications that GA is one of the most researched and implemented EOA. GA has innumerable variants and hybrids in the research domain. The terminology involved in GA shall be covered in later sections. However, it is necessary to again mention here that fit/fitness value/objective function value are to be considered as one and the same throughout the chapter. These terms shall determine which possible solutions from the pool of available solutions are better than the remaining solutions.

John Holland [136] describes the formulation of GA as an EOA on the principles of natural evolution.

DOI: 10.1201/b22647-3

3.2 TERMINOLOGY

The word "genetics" comes from the Greek word "genesis" which means "to grow". Let us explore some more biological terminologies which will come in handy for understanding GA.

1. **Chromosome:** The genetic information of any naturally living organism is stored in a chromosome. The chromosome itself is made up of several smaller parts called genes. A chromosome is represented in Fig 3.1

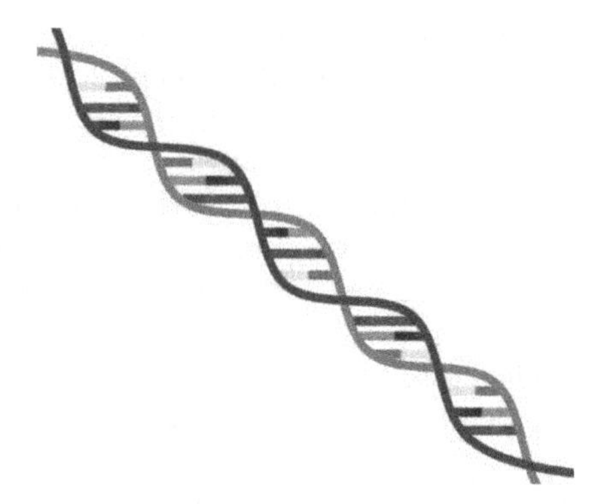

Figure 3.1: Chromosome

2. **Genes:** Genes are responsible to determine the characteristics of an individual. A gene represents a single independent feature/input variable. The relation between a chromosome and a gene is shown in Fig 3.2.

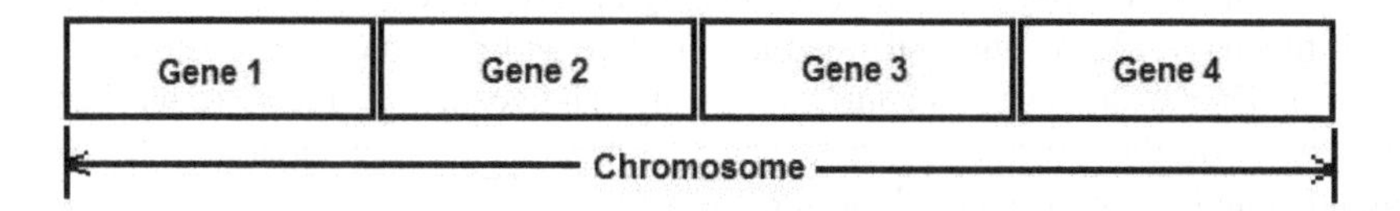

Figure 3.2: Genes and Chromosome

3. **Allele:** The possible values that a gene can have for a particular characteristic are called alleles. Like the gene for eye colour, the values can be black, brown, blue, etc.
4. **Pool:** The pool determines all the different possible combinations that the genes can be made up off. The more is the size of the gene pool, the better are the variations available for the newer generations.
5. **Genome:** The set of all the genes of a particular species in called genome. It is concerned more with the positioning of the genes. In biological terms, chromosome and genome may be different; however, in GA both are considered as the same.
6. **Mitosis:** In Mitosis, genetic information is copied from the parents to the offsprings.

7. **Meiosis:** Combination of different information from parents to form a new individual/offspring.
8. **Phenotype:** It is the term used to represent the objective function value.

In each iteration of a GA, the chromosomes or solutions have to go through the following processes (terms related specifically with GA):

1. **Selection:** Selection is used to select a certain number of chromosomes, as parents, from the mating pool (available solutions) for mating/reproduction.
2. **Crossover:** Crossover is a process to generate offsprings through the mating of parent chromosomes.
3. **Mutation:** Mutation is used to introduce variations in an existing chromosome.

There are different methods employed for the above processes, which will be discussed in later sections.

3.3 FUNDAMENTAL CONCEPT

GA is based on Darwin's theory of evolution as mentioned earlier in the chapter. Let us first understand how the Darwin's theory actually works and then apply it to form an optimization technique in this section.

Let us consider a herd of gazelles (a type of deer). The deer graze in open grasslands and are preyed upon by different predators. The deer have to sprint to save their lives. The ones that run faster are able to save their lives whereas the slower runners in the herd may fall prey to the predators. In other words, the gazelles that survive are the ones that can outrun at least one of the other gazelle in the heard. Thus the speed of sprinting plays an important role in the evolution of deer. The evolution of deer is presented in Fig 3.3.

A herd of deer is shown in Fig 3.3(a). These deer shall graze and move around in the grasslands. They are stalked and hunted down by predators. The original population shall be reduced to a smaller population through the killings done by predators. The remaining population is given in Fig 3.3(b).

The population shown in Fig 3.3(b) are the deer which were able to outrun their predators or peers. This population of deer is termed the selected population. All the deer in this population are better runners as compared to those deer which were hunted down by the predators.

Now this selected population or the remaining deer will have the chance to mate and reproduce. Thus the new population/generation of deer, i.e., the fawns, will have the genes of these deer. The new population is shown in Fig 3.3(c). The new fawns will have characteristics of the deer who could reproduce. The fawns will also contain certain mutations as a part of evolution. The process of selection of better candidates and their reproduction to form a new population is repeated again and again through the generations.

Nature has its own way of selecting individuals from the existing population. These surviving individuals then mate and reproduce a new generation. In the

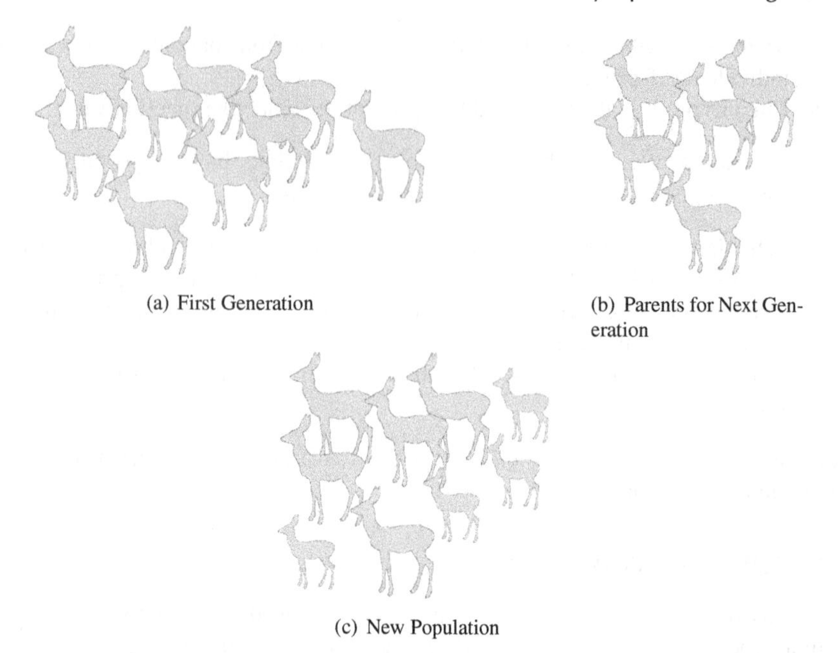

(a) First Generation

(b) Parents for Next Generation

(c) New Population

Figure 3.3: Concept of Genetic Algorithm

optimization process, of GA, this process of evolution is classified into three parts as: selection, crossover and mutation.

In GA, the chromosomes are randomly generated as the intial step. However, if these chromosomes are binary coded, which they should be in the actual version, then one needs to create the binary values for the input variables and form the chromosome. A simple example is presented for the same in the example section.

Once the population has been populated by the random chromosomes, the objective function for each chromosome is evaluated. This helps the process to pick the chromosomes to be used/selected for reproduction.

In the next phase of GA, the selection, crossover and mutation of chromosomes is performed to create new population. The selection, crossover and mutation can be done through different methods as mentioned in the following subsections.

3.3.1 SELECTION

Selection is a process to select the parent chromosomes from the existing population. These parent chromosomes will be reproducing the next generation of child chromosomes. The collection of fitter chromosomes which will become the parent chromosomes is termed the mating pool. There are a number of different methods applied for selecting the parent chromosomes. Usually the focus in selecting a parent chromosome is dependent on the objective function value of the individual chromosome. In this section, we will see the working of different selection methods which are implemented in GA. For this we assume a set of chromosomes which will be commonly referred to in all methods. Let the chromosomes in the mating pool have the objective function values as presented in the second column of Table 3.1.

Note: The construct of the chromosomes is not being discussed and therefore only the chromosome numbers are considered. The construct of the chromosomes shall be demonstrated in the Example section. For the same reason the objective function value is not presented here.

The selection methods usually work good or can be explained well with maximization problem. Thus the objective function values for the same chromosome set converted to maximization problem (1/(Obj Fun Value)) is given in the last column. There are a large variety of methods used for selecting chromosomes in the mating pool, some of which are explained below:

Table 3.1
Existing Population

Chromosome Number	Objective Function Value (Min)	Objective Function Value (Max)
1	0.0556	18
2	0.125	8
3	0.0256	39
4	0.0476	21
5	0.0714	14

1. **Roulette Wheel Selection**: In Roulette Wheel Selection method, each eligible chromosome is assigned with a probability. The probability of the ith chromosome is evaluated using Eq 3.1

$$p(i) = \frac{f(i)}{\sum_{j=1}^{n} f(j)} \tag{3.1}$$

where, f(i) is the objective function value of the ith chromosome and 'n' represents number of chromosomes available for selection.

It is better to apply maximization objective function values, so that the chromosome with better, i.e., minimum objective function value, gets maximum value, range and space in Roulette Wheel Selection method. The probabilities of the chromosomes mentioned in Table 3.1 (column 3, maximization problem), are evaluated and presented in Table 3.2. In Table 3.2, the sum of objective functions is calculated and presented in the last row. In this case the sum is found to be 100. Then, by dividing all the objective function values by this sum, the probability of each chromosome is obtained. The sum of objective function values may differ, but the sum of probabilities will always be 1.

Based on the above probabilities a Roulette wheel is formed. The chromosomes get their allocation in the Roulette wheel based on their objective

Table 3.2
Roulette Wheel Selection

Chromosome Number	Obj Fn Value	Probability
1	18	0.18
2	8	0.08
3	39	0.39
4	21	0.21
5	14	0.14
Sum	100	1

function value/probability. The values presented in the Roulette wheel are in percentages, i.e., out of 100.
The Roulette wheel for the data in Table 3.2, is given in Fig 3.4.

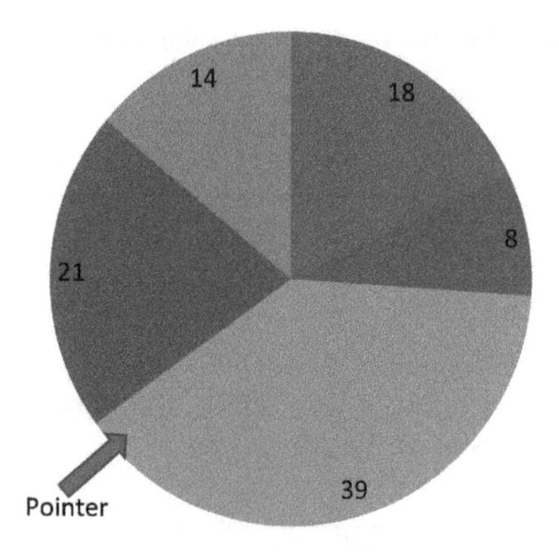

Figure 3.4: Real Roulette Wheel

This is a physical Roulette wheel, however, it will not be helpful in selecting chromosomes for the mating pool. Thus some reconfiguration is required. The allocation of Roulette wheel can be practically represented as in Fig 3.5. With this representation, the probabilities of all the chromosomes are spread through specific ranges of values [12]. For chromosome 1 the range of values are from 0 to 0.18, while for chromosome 2 the range is from 0.18 to 0.26. For chromosome 3, the range of numbers is from 0.26 to 0.65 with a difference of 0.39 and so on as shown in figure. This distributes the ranges amongst the chromosomes based on their probabilities.
A random number between 0 and 1 is generated and the chromosome in whose range the random number lies is selected in the mating pool.

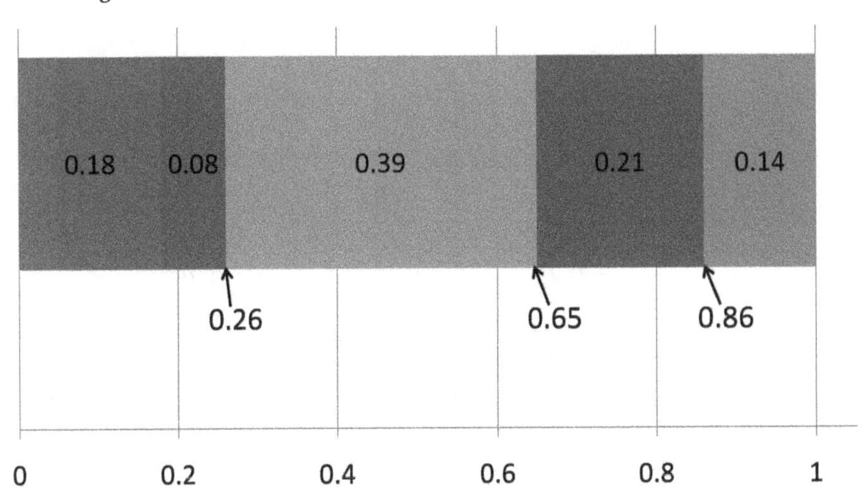

Figure 3.5: Roulette Wheel Implementation

There are multiple variations in Roulette Wheel Selection method. In one of the variations, the probability is not found and the random number is generated between 0 and $\sum_{j=1}^{n} f(j)$.
Roulette Wheel Selection method has the negative effect of early convergence, since the child chromosomes are concentrated near the chromsomes having larger ranges.

2. **Stochastic Universal Sampling**: In some references, the straight line implementation of Roulette wheel, as shown through Fig 3.5, is itself considered as a Stochastic Universal Sampling. However, another version places markers at specific intervals to select the parent chromosomes in the mating pool.
For example, a marker can be placed exactly opposite to the random number. If the random number $r > 0.5$, then the marker would be at 'r−0.5' and if $r < 0.5$, then the marker can be placed at 'r+0.5'. Generation of one random number is thus used to select more number of parents through these markers. In Stochastic Universal Sampling, one parent is selected through the random number and the remaining through markers. Fig 3.6 shows an example of Stochastic Universal Sampling. In the figure, 'r' represents the random number generated and 'r' represents the marker. If $r < 0.5$, marker is placed at 'r+0.5' and if $r > 0.5$, marker is placed at 'r−0.5'.
More than one markers can also be placed in relation to the random number, if required. This method helps to oversome the negative effects mentioned in the Roulette Wheel Selection method.
3. **Rank Selection**: In rank selection method, the chromosomes eligible to enter the mating pool are arranged based on their objective function value. The chromosome with the best objective function value gets the highest rank while the worst chromosome gets rank 1. Highest rank is equal to the number of

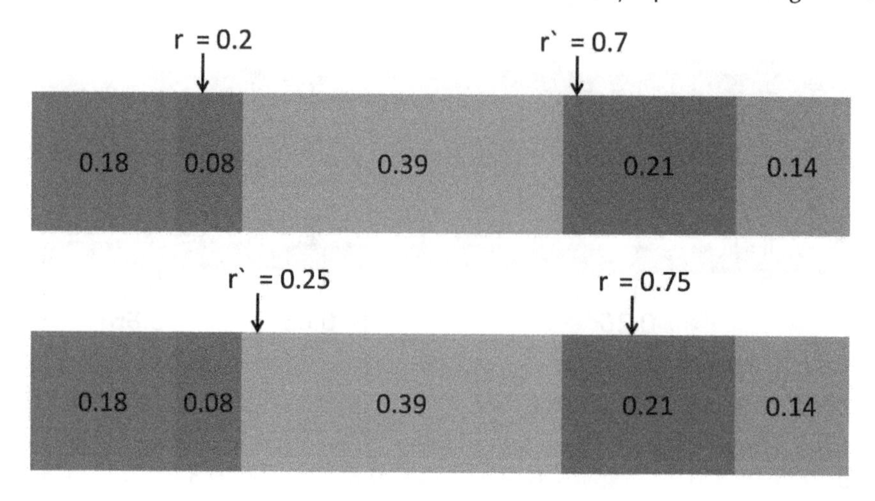

Figure 3.6: Stochastic Universal Smapling

chromosomes eligible to enter the mating pool, let us assume it to be 'n'. The probability of each chromosome is evaluated as given in Eq 3.2 [47].

$$p(i) = \frac{rank(i)}{n \times (n-1)} \tag{3.2}$$

The probabilities based on the above equation is presented in Table 3.3

Table 3.3
Rank Selection Method

Chromosome Number	Obj. Fn. Value	Rank	Probability
3	0.025641026	5	0.25
4	0.047619048	4	0.2
1	0.055555556	3	0.15
5	0.071428571	2	0.1
2	0.125	1	0.05

The chromosomes are arranged in ascending order based in objective function values. There is no need to convert the minimization problem into a maximization problem. The ranks shall remain the same even when the chromosomes change, i.e., the first ranked chromosome will always have a probability of 0.25 and the least ranked chromosome will always have 0.05, and that applies to all the other chromosomes in the mating pool.

Once the probabilities are associated with the chromosomes, then random chromosomes can be selected in a similar way as in the Roulette Wheel selection method.

The spread of chromosomes is presented in Fig 3.7.

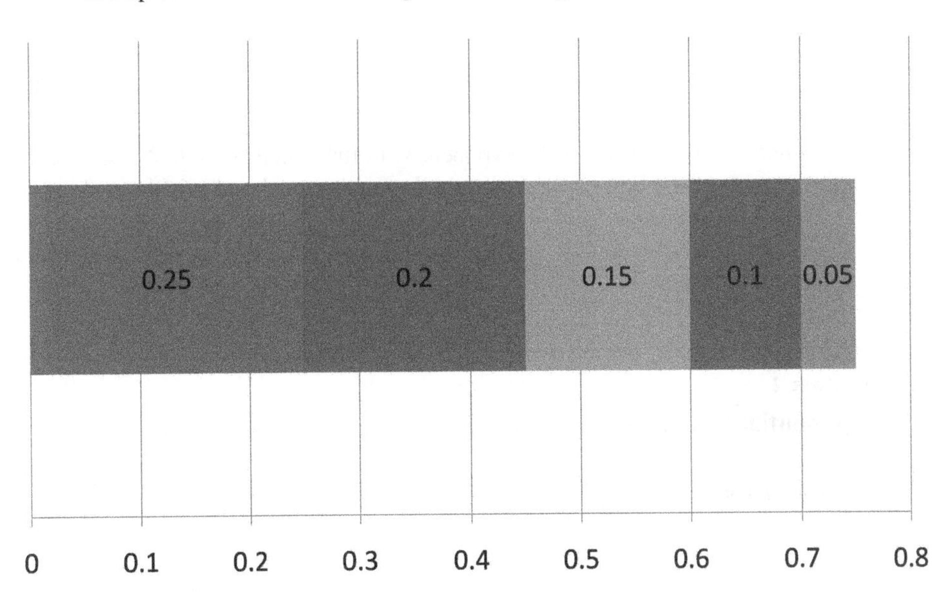

Figure 3.7: Rank Selection

Here the random number can vary from 0 – 0.75, as the sum of all the probabilities is 0.75. The range of first chromosome is from 0 to 0.25, for second it will be from 0.26 to 0.45 and so on.

Other than this method, there are some more variants in Rank Selection method. One of the variant is termed **Linear Rank Selection**. The equation used in this method is given by Eq 3.3 [50].

$$p(i) = \frac{1}{N}\left(n^- + (n^+ - n^-)\frac{i-1}{N-1}\right) \tag{3.3}$$

where $\frac{n^-}{N}$ gives the ratio of worst chromosome (n^-) to the population size (N) whereas $\frac{n^+}{N}$ gives the ratio of best chromosome (n^+) to the population size (N) and 'i' is the chromosome whose probability is being evaluated.

The above equation does not require the objective function value of the chromosome as the probabilities are evenly distributed.

Let us evaluate the calculation for a mating having 5 chromosomes.

$p(1) = \frac{1}{5}\left(1 + (5-1)\frac{1-1}{5-1}\right) = 0.2$

$p(2) = \frac{1}{5}\left(1 + (5-1)\frac{2-1}{5-1}\right) = 0.4$

Thus, as the name suggests, there is even distribution of probability amongst the chromosomes. Probability for more number of chromosomes can be evaluated by the readers.

Another variant in Rank selection method is the **Exponential Rank Selection**. The chromosomes do not get linearly distributed probabilities but are exponentially weighted [16] in this method. The probability distribution for exponential rank selection is given in Eq 3.4

$$p(i) = \frac{c^{N-i}}{\sum_{j=1}^{N} c^{N-j}} \tag{3.4}$$

where 'c' is the base of the exponent with the condition of $0 < c < 1$, 'N' is the rank of the best chromosome in the mating pool and 1 is the rank of the worst chromosome.

Table 3.4 gives a sample calculation of the probabilities through Exponential Rank Selection. In the above table, the rank of the chromosomes are given in

Table 3.4
Exponential Rank Selection Method

Chromosome Rank	c=0.5		c=0.99		c=0.01	
	c^{N-j}	p(i)	c^{N-j}	p(i)	c^{N-j}	p(i)
5	1	0.516129	1	0.20404	1	0.99
4	0.5	0.258065	0.99	0.202	0.01	0.0099
3	0.25	0.129032	0.9801	0.19998	0.0001	9.9E-05
2	0.125	0.064516	0.970299	0.19798	0.000001	9.9E-07
1	0.0625	0.032258	0.960596	0.196	1E-08	9.9E-09
Sum	1.9375	1	4.900995	1	1.010101	1

the first column. We have assumed that the mating pool has 5 chromosomes. The probabilities for these 5 chromosomes is evaluated when c= 0.5, c=0.99 and c=0.01, so that the extreme and middle conditions can be visualized. The $\sum_{j=1}^{N} c^{N-j}$ is given at the bottom of the table for each column and respective value of 'c'. The c^{N-i} is given for each chromosome in their respective rows. The probabilities are so distributed that the sum of probabilities is always 1. Fig 3.8 gives a visual representation for the three case presented in Table 3.4. The figure indicates that there is an equal distribution of probability for c = 0.99 whereas a highly biased probability distribution amongst the chromosomes for c = 0.01.

The merit of Rank selection methods is that it is more robust and all the chromosomes have a chance to get selected as a parent.

However, the downside associated with these methods is that in some cases it is unfair to chromosomes with better objective function values. The probabilities associated with chromosomes is not at all related with their objective function value and can lead to slower convergence in the optimization process.

4. **Tournament Selection**: One of the most popular methods for selecting parent chromosomes is the Tournament Selection method. In this method, some chromosomes are selected randomly from the mating pool and are made to compete against each other [47]. The competition is based on the objective function values of the selected parent chromosomes. The winner of the tournament gets to

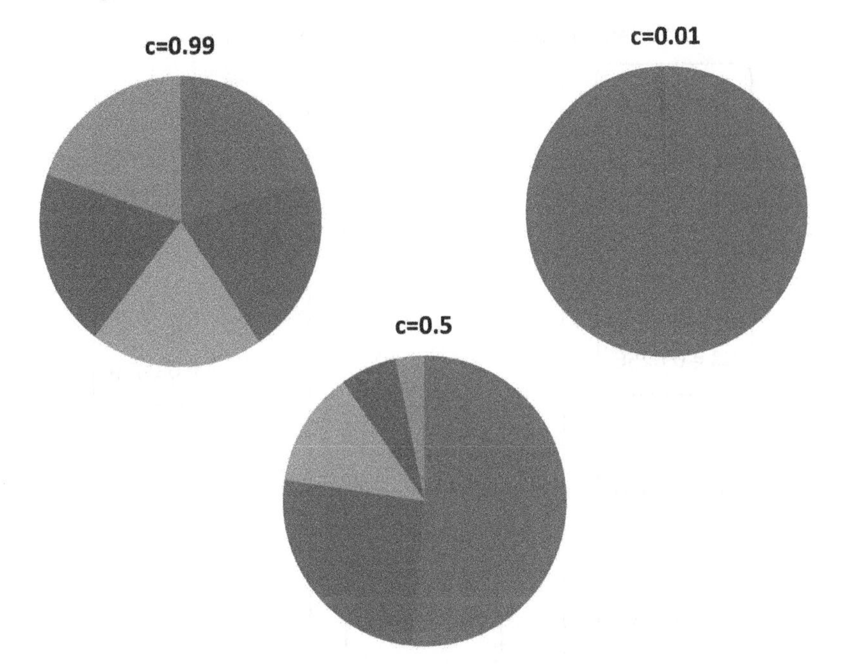

Figure 3.8: Exponential Rank Selection

reproduce. The tournament size represents the number of chromosomes participating in the tournament. In binary tournament, only two chromosomes participate whereas multiple chromosomes participate in large tournaments. Tournament size is very important, as a large tournament size will lead to a loss of diversity. Fig 3.9 shows the process of chromosome selection in Tournament Selection. As can be seen from the figure, it is not necessary that the best chromosome will be a part of the tournament. Initially, in this case, three chromosomes are selected randomly and then the best chromosome from the randomly selected chromosomes is picked for reproduction.
Some of the merits of Tournament Selection are that it has less time complexity, parallel implementation of the method can be achieved, not necessarily dominated by better individuals and no scaling or sorting is required.

5. **Boltzmann Selection**: This selection method is inspired from Simulated Annealing. The diversity is maintained in selection so as to achieve the Boltzmann distribution. The probabilities of the chromosomes change as the iterations increase which is similar to the temperature decrease in Simulated Annealing. The probability for a parent chromosome in Boltzmann Selection is given by Eq 3.5.

$$p(i) = exp\left[-\frac{f_{max} - f(x_i)}{T}\right] \tag{3.5}$$

where p(i) is the probability of chromosome ‘i’, f_{max} is the objective function

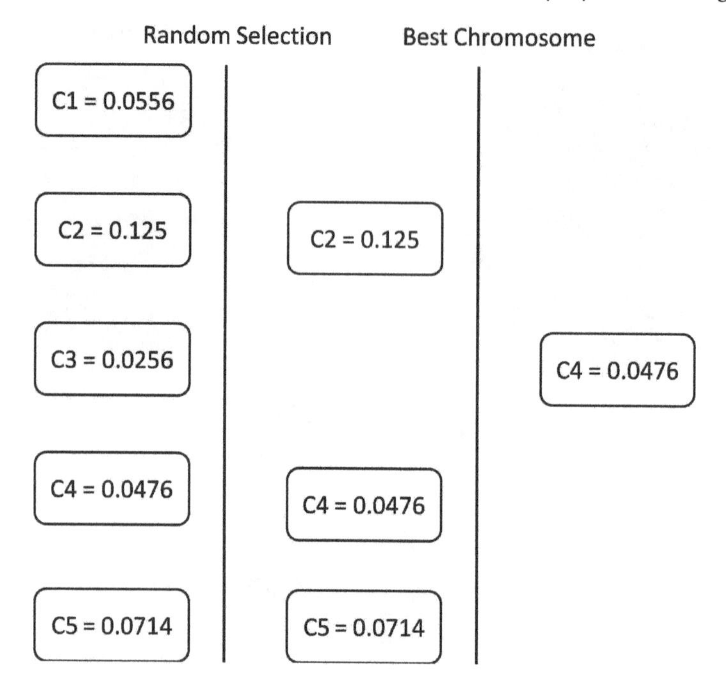

Figure 3.9: Tournament Selection

value of the best chromosome, $f(x_i)$ is the objective function value of chromosome 'i' and $T = T_0(1-\alpha)^k$ in which α is to be chosen in between 0 and 1, whereas T_0 lies between 5 and 100 and k = (1+100*g/G), where 'g' is the current iteration and 'G' is the maximum number of iterations.

A simple calculation for probabilities using Boltzmann Selection is presented in Table 3.5. In Table 3.5, the columns are as: the first column gives the chromosome number, the second column gives its objective function value, the third column gives the best objective function value (as we are considering maximization problem) and the fourth column gives the difference between the best chromosome and the respective chromosome. Then in the next sections the probability calculation is given for parameters expressed at the bottom of the table. In the first section, g = 1 and in the next section g = 99 while other parameters remain constant. Thus we can observe, from Table 3.5, that the probability distribution in the beginning of the search process (i.e., when g = 1), is distributed, whereas at the end (i.e., when g = 99), only the best chromosome is selected for reproduction.

Figure 3.10 represents the probability distribution in Boltzmann Selection method. The even distribution of probabilities at the beginning of the search process can be observed (at g = 1), while at the end only the best chromosome can be selected (at g = 99).

Table 3.5
Boltzmann Selection Method

Chromosome No	Obj Fn	Best Obj Fn	$f_{max} - f_i$	k	T	p(i)	k	T	p(i)
1	18	39	21	2	64.8	0.7232	100	0.0021	0
2	8	39	31	2	64.8	0.6198	100	0.0021	0
3	39	39	0	2	64.8	1	100	0.0021	1
4	21	39	18	2	64.8	0.7575	100	0.0021	0
5	14	39	25	2	64.8	0.6799	100	0.0021	0
				T_0		80	T_0		80
				α		0.1	α		0.1
				g		1	**g**		99
				G		100	**G**		100

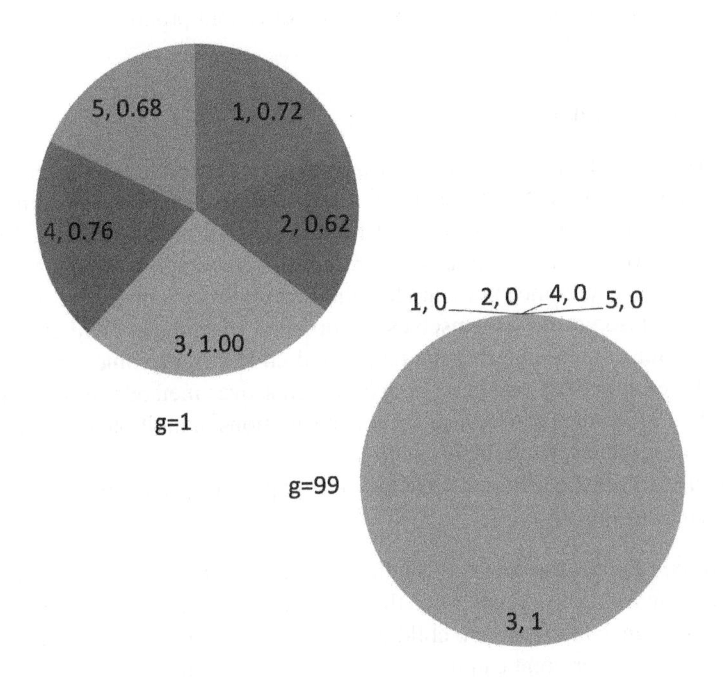

Figure 3.10: Boltzmann Selection

6. **Others**: There are some more selection methods implemented for populating the mating pool. These methods are enlisted below:
   a. **Elitism**: This is a supportive process to help preserve the best chromosomes which may be lost in the process of Crossover and Mutation. Through Elitism, the best chromosome (or some chromosomes) may be copied 'as it is' in the next generation of population. A drawback of Elitism may be the dominance of the best chromosome/local optima in the search process.
   b. **Age Based Selection**: In this process of selection, the chromosomes are allowed to stay in the race of reproduction only up to a certain prespecified number of iterations, i.e., age. After completion of these prespecified number of iterations, the chromosomes shall be shunted even though they may have better objective function value.
   c. **Truncation Selection**: This method is utilized for large number of selections or mass population. In this method the chromosomes are sorted based on their objective function values and only the best chromosomes are selected [53]. A threshold is defined through Trunc, having values from 10% to 50%, below which the chromosomes are not allowed to reproduce. In other words, the top 10% to 50% from the sorted list of parent chromosomes are only selected while the remaining are truncated. The chromosomes above the specified threshold produce offspring in a uniform order.

3.3.2 CROSSOVER

The selection methods described above are used for selecting chromosomes for mating. Crossover is used to combine the features of two or more chromosomes to produce the child chromosomes, through the usage of different methods. Before we move to the working of the process of crossover, some important points to be noted are: 1) the parent chromosomes and the child chromosomes may be different, or the parent chromosomes may themselves be modified to generate child chromosomes. In this section, we assume that the parent and child chromosomes are different. 2) GA may be real coded or binary coded and crossover methods may be applied as applicable. 3) Readers can themselves make variations in multiple methods or create a hybrid to suit their requirements.

Let us explore the different crossover techniques for producing the next generation of chromosomes:

1. **Single Point Crossover**: In single point crossover, the parent chromsomes are bisected at a given point. Then the corresponding parts of the chromosomes are interchanged to form two child chromosomes from two parent chromosomes. The crossover point can be: fixed, randomly chosen, based on some function, or any other suitable technique may be applied.
   The Single Point CrossOver is given in Fig 3.11 with (a) real coded GA and (b) binary coded GA. The vertical dashed line shows the point of crossover. It

can be seen that for child 1, the part of chromosome before the crossover point is taken from parent 1 and the remaining part is taken after the crossover point from parent 2. For child 2, the portions taken from the parents are reversed i.e, part of the chromosome after the crossover point from parent 1 and before the crossover point from parent 2.

There is also a possibility of generating a single child instead of two. In other words, one of the child chromosomes may not be generated at all.

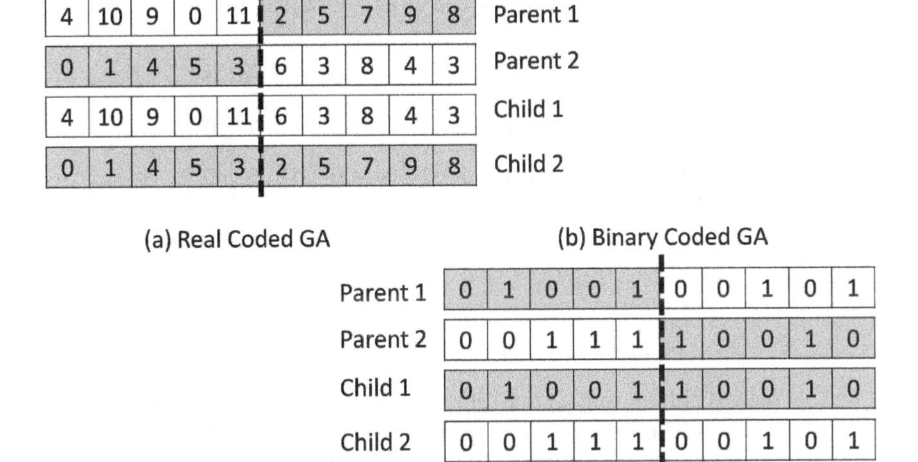

Figure 3.11: Single Point Crossover

2. **Two Point Crossover**: A two point crossover is similar to a single point crossover method, except that there are two points of crossover. The method of reproduction through Two Point Crossover is presented in Fig 3.12. As can be

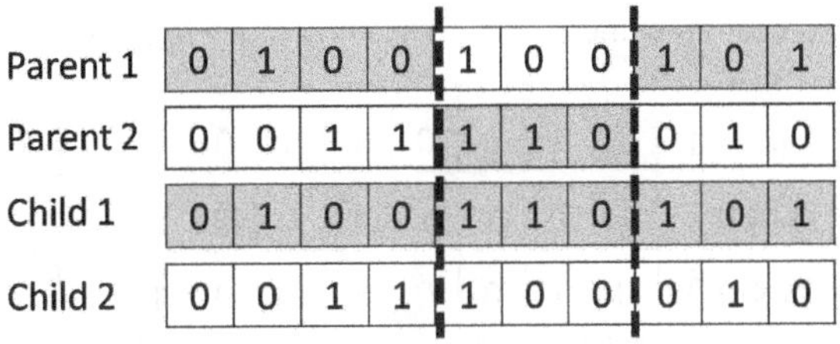

Parent 1	0	1	0	0	1	0	0	1	0	1
Parent 2	0	0	1	1	1	1	0	0	1	0
Child 1	0	1	0	0	1	1	0	1	0	1
Child 2	0	0	1	1	1	0	0	0	1	0

Figure 3.12: Two Point Crossover

seen from the figure, the child chromosomes contain contents from different parents on either side of the crossover points. As presented before in one point crossover, in two point crossover, the crossover points can be located on the basis of different methods.

The downside of having more numbers of crossover points is that the combination blocks may be disrupted, whereas a merit of having more numbers of crossover points is increased in exploration of the search method.
As an extension to the two point crossover method, the number of crossover points can further be increased in a multi point crossover.

3. **Uniform/Masked Crossover**: Uniform crossover introduces a masking chromosome in the process. A binary chromosome having a length similar to the parent chromosomes is created. Fig 3.13 shows the crossover of parents through a masked chromosome. It can be observed that for the locations where

Parent 1	0	1	0	0	1	0	0	1	0	1
Parent 2	0	0	1	1	1	1	0	0	1	0
Mask	0	1	0	0	1	1	0	1	1	0
Child 1	0	1	1	1	1	0	0	1	0	0
Child 2	0	0	0	0	1	1	0	0	1	1

Figure 3.13: Uniform/Masked Crossover

the masked chromosome has a 1, the Child 1 gets the value from Parent 1 and the locations where the masked chromosome has a 0, the Child 1 gets the value from Parent 2. On the other hand, Child 2 gets data from Parent 1 for the locations where masked chromosome has a 0 and from Parent 2 where the masked chromosome has a value 1.

4. **Three Parent Crossover**: This is a unique kind of a crossover as compared to the methods studied until now, for the reason that it involves three parents for reproduction and the number of childs produced is only one. The methodology for producing the child is that if the values of Parent 1 and 2 are the same for a particular location, then that value is copied in the Child, otherwise the value of the third chromosome is copied as shown in Fig 3.14.

Parent 1	0	1	0	0	1	0	0	1	0	1
Parent 2	0	0	1	1	1	1	0	0	0	0
Parent 3	0	1	0	0	1	1	0	1	1	0
Child	0	1	0	0	1	1	0	1	0	0

Figure 3.14: Three Parent Crossover

5. **Partially Matched Crossover**: The Partially Matched Crossover, often termed PMX, is a special crossover. It is applied in problems similar to the Traveling Salesman Problem (TSP). As seen in earlier chapters, the speciality of TSP is that the values cannot be repeated in a chromosome, as the salesman can visit

any city only once and a value represents the particular city. Let us explore how PMX works. The parent chromosomes are taken as in Fig 3.15:

Parent 1	3	6	2	7	1	5	9	4	8
Parent 2	5	7	9	8	2	1	4	3	6

Figure 3.15: Parents for Partially Matched Crossover

Each parent chromosome consists of 9 values. As can be observed, the numbers in each cell are distinct and are not repeated. Let us assume that there are two crossover points after third and seventh positions.
In PMX, to create the first child, the values between the two crossover points from parent 1 is first copied in the respective positions of the child chromosome. Next it is required to know as to which values are present in between the crossover points in parent 2 as well. This can be understood from Fig 3.16. It

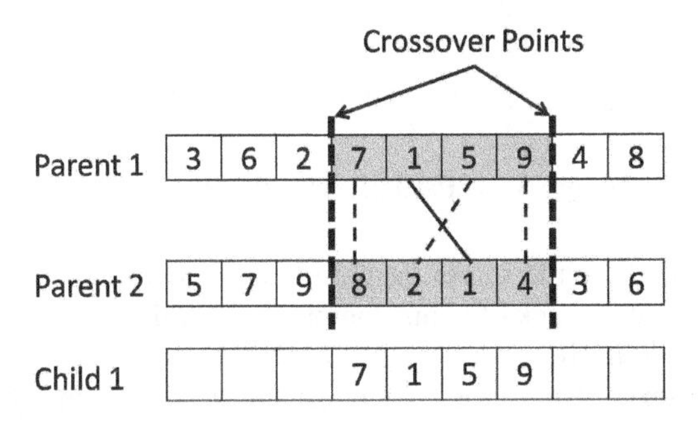

Figure 3.16: Partially Matched Crossover - Child 1 - Step 1

can be seen that '1' is present in both parents 1 and 2 in between the crossover points. The uncommon values are 7, 5 and 9 from parent 1. The uncommon values shall be plotted in between parents 1 and 2. The solid lines in the figure represents the common values whereas the dashed lines represents and plots the uncommon values between parents 1 and 2.
Thus the plotting of uncommon values between parents 1 and 2 are as:

$$7 \leftrightarrow 8, 5 \leftrightarrow 2, 9 \leftrightarrow 4$$

Let us refer Fig 3.16, as the contents between the two crossover points have been copied from parent 1 into child 1, the remaining contents should be copied from parent 2. However, if the remaining contents are copied as it is from parent 2, then values 7, 5 and 9 shall be repeated. Repetition of values is however not allowed.
In PMX, therefore a slightly different approach is followed. If we go by the above method, it is obvious that 7, 5 and 9 will be repeated, but on the other hand 8, 2 and 4 will not be a part of the child chromosome. Thus, plotting of values is done, as shown in Fig 3.16. It can be seen that 8 is plotted for

7, and thus 8 shall take the place of 7 in the values being copied from parent 2 to the child 1 chromosome. Similarly, 5 will be replaced by 2 and 9 by 4. In this way, the values shall not be repeated in the child chromosomes. The process is explained in Fig 3.17. The remaining values shall be copied in the child 1 chromosome in their original position from parent 2. A similar method shall be followed for generating the second child with the roles of parents 1 and 2 being interchanged. This method is of specific use and should be used

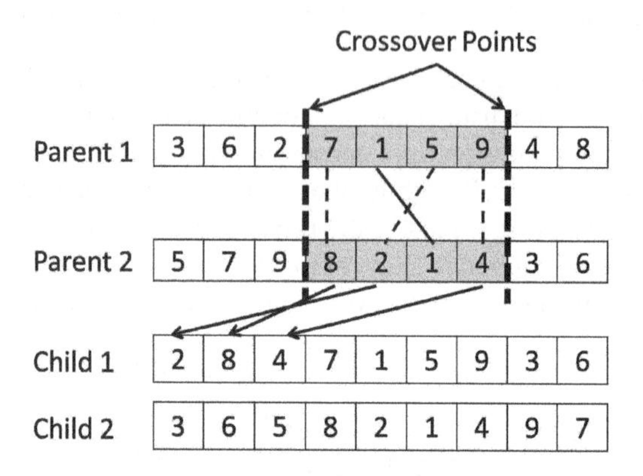

Figure 3.17: Partially Matched Crossover

in problems where the values cannot be repeated within a chromosome. The method has a drawback of being complex but the disadvantage is the property of the problem itself.

6. **Ring Crossover**: Ring crossover is usually not found in references but is a very easy method to perform crossover. In a ring crossover, the parent chromosomes are connected to each other to form a ring and then they are cut diagonally to form two child chromosomes as shown in Fig 3.18. In Ring crossover, the method of connecting the parent chromosomes and generating the child chromosomes can be varied, like one parent can be concatenated at the end of another parent. Also the placement of the diagonal can be varied to generate different types of results. As can be seen from the process, the effectiveness of the Ring Crossover is dependent on a number of factors. It is also important to visualize that the programming shall be a little tricky for this crossover.

There are many other crossover methods which may be applied to specific conditions or in general as well like, Precedence Preservative Crossover, Ordered Crossover, Shuffle Crossover, Reduced Surrogate Crossover, etc.

3.3.3 MUTATION

The process of selection selects the parents. The process of crossover produces child chromosomes. The next process in evolution is mutation. In mutation, the chromosome is changed in an asexual manner. Mutation is helpful in adding newer elements

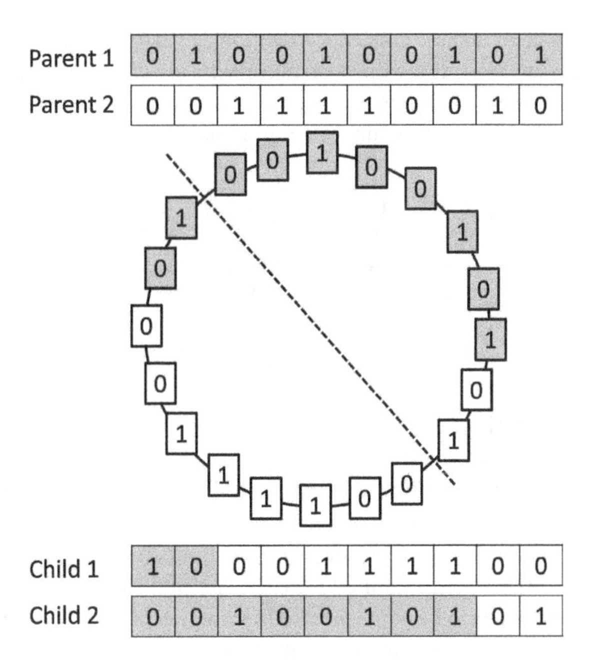

Figure 3.18: Ring Crossover

in the chromosome pool, which in turn avoids the process getting trapped in a local minima. Even though there is no specific classification of mutation, let us enlist some of the common methods. Mutation in itself can also be used for producing child chromosomes.

1. **Binary Mutation**: As the name implies, this type of mutation is applied on a chromosome having binary values. In binary mutation, mutation happens through the flipping of individual bits, i.e., 1 is replaced with a 0 and vice versa. The bits which are to be flipped are chosen through probability generation. The probability of each bit to be flipped is generally taken to be inverse of the length of the chromosome. A simple example of binary mutation is shown in Fig 3.19.

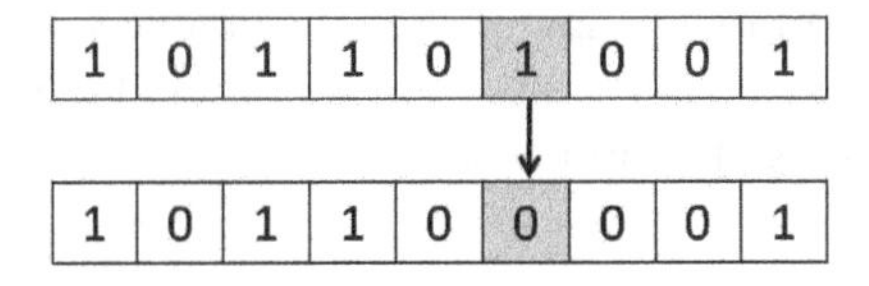

Figure 3.19: Binary Mutation

2. **Interchanging Mutation**: As discussed in the previous section, problems like traveling salesman, requires that all the values should be present in the

chromosome. In such a case, the mutation is performed through Interchanging. In this method, the values of two positions within the chromosome are interchanged. Interchanging mutation can be applied to a normal chromosome as well. Fig 3.20 shows the interchanging mutation.

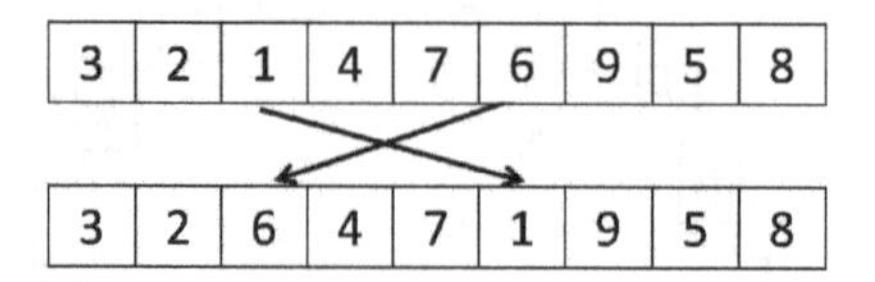

Figure 3.20: Interchanging Mutation

3. **Uniform Mutation**: This mutation is applied to real coded GA. In Uniform mutation, a random position is selected and its value is replaced with a random number which lies between the allowable limits of the input variable. When applied to multiple elements, it is termed multiple uniform mutation.
4. **Gaussian Mutation**: In Gaussian mutation, all the elements in the chromosome are mutated based on the Gaussian Distribution function. The Gaussian Distribution function has a mean of zero and the variance is adaptive. The effect of mutation reduces as the iterations increase. The normal distribution is given by $N(\mu,\sigma^2)$, where μ is 0 and σ is evaluated as: $\sigma_k = \frac{T-t}{T}.\frac{v_k^{max}-v_k^{min}}{3}$ where k is the position of the element, T is the total number of iterations, t is the current iteration number, v_k^{max} and v_k^{min} are the limits of the input variable [18].

After completion of selection, crossover and mutation, the objective function values of the new population is evaluated. The process is iterated through a number of times until the stopping criterion is satisfied.

3.4 ALGORITHM AND PSEUDOCODE

A simple algorithm for GA is presented in Alg 3.1. The steps are aggregated for a better understanding. The optimization process starts with the generation of random chromosomes within the population. After initializing the random chromosomes, the iteration process starts. The first step is to obtain the objective function value of each chromosome. Once this is done we move for creation of new chromosomes. The parents are selected through any of the selection processes, then crossover is performed and lastly mutation. The new population of chromosomes, will now be used to replace the existing chromosomes. The process is repeated until the stopping criterion is met.

Algorithm 3.1 Genetic Algorithm

```
START
Initialize random chromosomes in a population
for 1 to Maximum Number of Iterations do
   for 1 to Population do
      Find Objective Function value for respective chromosome
   end for
   for 1 to N do
      Perform Selection process
      Perform Crossover
      Perform Mutation
   end for
   Replace current population with new population
end for
Display Result
STOP
```

In the above algorithm, it is assumed that the stopping criterion is based on the number of iterations. Also, selection process shall itself be requiring an algorithm of its own, however, due to the multiplicity of methods, it has been represented through a single command. The crossover process, is based on probability and these random numbers will have to be generated at run time. Similar is the case with mutation, where random numbers will have to be generated. Lastly, the whole of the population may not be replaced. Hence the command should be assumed to replace a part of the population with the new chromosomes. Also, from the second iteration onwards, the objective function value of the new chromosomes will only be evaluated.

Alg 3.1, is a very simple representation of GA through an algorithm. The algorithm shall drastically be changed based on the methods implemented along with the factor of the optimization process being binary coded or real coded. In case of binary coded GA, the binary chromosomes shall have to be decoded each time a new chromosome is generated for evaluation of its objective function value.

3.5 FLOWCHART

In Fig 3.21, a simple representation of the flowchart for GA is displayed. The GA process starts with the generation of random population of chromosomes. These chromosomes should be spread throughout the search space for better results.

After the generation of random chromosomes, the objective function values of these chromosomes are evaluated. For this, we need to decode the actual values of input variables from the encoded chromosome strings. Once the objective function values of the chromosomes are evaluated, then the next step of selection can be performed. In selection of chromosomes, any one of the different techniques like roulette wheel, tournament selection, etc can be used. The selection of chromosomes leads to crossover for production of new population. Again the crossover can be

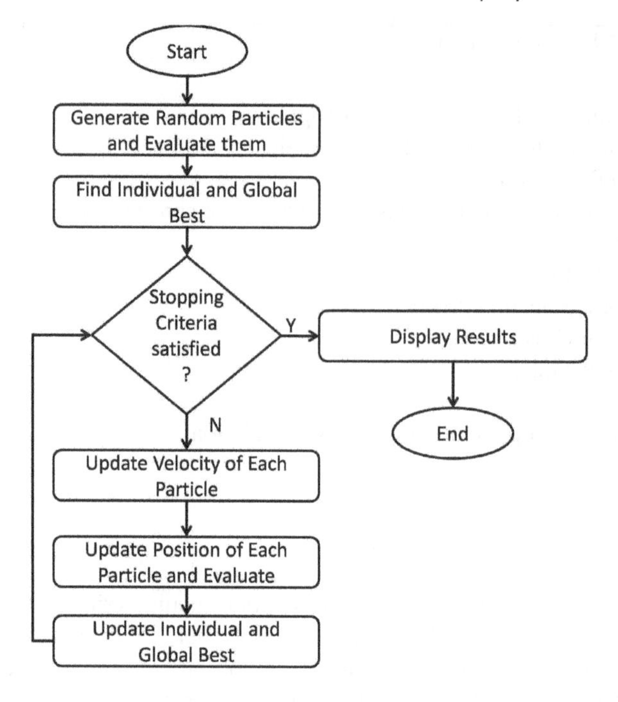

Figure 3.21: Flowchart

selected as single point or any other method as suited for the optimization problem in hand. After completion of crossover, mutation is performed on the new population of chromosomes. Mutation leads us to the end of the iteration. The stopping criterion needs to be checked. If the stopping criterion is not met, the optimization process moves towards the next iteration. The objective function value of the new chromosomes are evaluated and the cycle continues. On completion of the process, i.e., stopping criterion being achieved, the results are displayed and the optimization process is exited.

3.6 EXAMPLE

GA is an EOA and thus unlike the traditional search methods does not work with a single solution. It creates a number of possible solutions for searching the optimal solution. GA has got a large amount of variations in its process itself. In the first step of GA, we need to generate a collection of possible solutions. One solution shall be represented by a chromosome and one factor/input variable shall be represented by a gene. The chromosome that we initially consider will be made of binary string as shown below in Fig 3.22:

This is a binary representation of a chromosome. The length of each gene is arbitrary and depends on the step size of an input variable along with its range. The gene should be able to represent the lower and upper bounds of the input variable

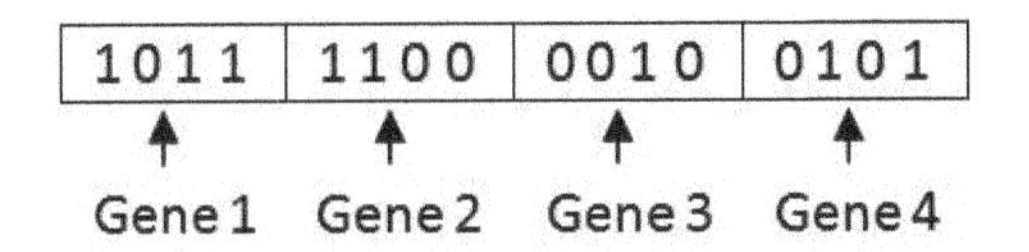

Figure 3.22: Representation of Chromosome

along with acceptable step size. A gene having a length of ‘n’ can represent $2^n - 1$ intervals and 2^n values. The size of each interval will be given as: $\frac{range}{2^n-1}$. However, binary representation of chromosome has its own drawbacks and can cause a lot of problems specially in encoding and decoding the values and for input variables which have continuous values.

Let us now solve the Griewank Equation (Eqs 2.5) using GA. Eqs 2.9 and 5.9 can be referred for details about solving the Griewank Equation. The input variables are assumed to be 5. The length of one gene is taken as 4 digits each. The step size would be very high and the process will never be able to find the optimal solution. It is therefore suggested that the length of the gene should be increased for better resolution and accuracy.

One chromosome can now be represented as given in Table 3.6.

Table 3.6
One Chromosome having Five Genes

	Gene 1	Gene 2	Gene 3	Gene 4	Gene 5
Sample Chromosome	1 0 1 1	0 1 1 0	0 1 0 1	0 0 1 0	1 1 1 0

However, now the bigger issue is how to decode the chromosome? What do we understand from the above data? Thus let us go back a little and remind ourselves that the gene should represent the lower and upper bound of an input variable. In this case the lower and upper bounds are −600 and 600, respectively. Each interval is given as $\frac{range}{2^n-1}$. The total range is 1200 and $2^n - 1$ stands for 15, thus one interval size comes out to be 80. The genetic values can thus be decoded as given in Table 3.7:

Thus mapping the gene values in the sample chromosome from Table 3.6 to Table 3.7, the gene values are found to be: 280, −120, −200, −440, 520.

Table 3.8 gives a population of six chromosomes which will be used for further calculations and their actual values. The genes in a chromosome should be generated randomly.

It is very clear then that the chromosomes may be generated randomly based on the string size, but to obtain the objective function value, we need to decode their values.

Now that we know the method to decode and encode the chromosomes, let us evaluate the objective function values for the chromosomes. These objective function values are evaluated for the Griewank function and listed in Table 3.9.

Table 3.7
Genes and Their Corresponding Values

Gene	Value
0 0 0 0	–600
0 0 0 1	–520
0 0 1 0	–440
0 0 1 1	–360
0 1 0 0	–280
0 1 0 1	–200
0 1 1 0	–120
0 1 1 1	–40
1 0 0 0	40
1 0 0 1	120
1 0 1 0	200
1 0 1 1	280
1 1 0 0	360
1 1 0 1	440
1 1 1 0	520
1 1 1 1	600

Table 3.8
Population of Chromosomes

Population	Genes	Actual Values
Chromosome 1	1 0 1 1—0 1 1 0—0 1 0 1—0 0 1 0—1 1 1 0	280 –120 –200 –440 520
Chromosome 2	1 1 1 1—1 0 1 0—0 1 0 1—0 0 1 1—0 1 1 0	600 200 –200 –360 –120
Chromosome 3	0 0 0 1—0 0 1 0—0 0 0 0—1 0 1 0—1 1 0 0	–520 –440 –600 200 360
Chromosome 4	1 0 0 1—0 1 1 1—0 0 0 0—0 1 1 0—1 1 0 0	120 –40 –600 –120 360
Chromosome 5	1 1 1 1—1 1 1 0—0 1 1 1—1 0 1 1—0 0 1 1	600 520 –40 280 –360
Chromosome 6	1 0 0 1—1 0 0 0—1 0 0 1—0 0 0 1—1 0 0 0	120 40 120 –520 40

Table 3.9
Objective Function Values of Initial Chromosomes

Population	Actual Values	Obj. Fn. Value
Chromosome 1	280 –120 –200 –440 520	150.8578
Chromosome 2	600 200 –200 –360 –120	147.4149
Chromosome 3	–520 –440 –600 200 360	249.3727
Chromosome 4	120 –40 –600 –120 360	131.3716
Chromosome 5	600 520 –40 280 –360	211.0631
Chromosome 6	120 40 120 –520 40	76.2643

Once the objective function values are evaluated for the chromosomes, we can proceed towards the first part of the iteration process in GA, i.e., selection of chromosomes for the next iteration.

We shall inculcate two methods of selection first, truncation with a threshold of 50% and second, Roulette Wheel Selection method. A lesser limit for threshold could have been selected, however, due to smaller population size, 50% is taken. Also we have selected Roulette Wheel for the first iteration and in the second iteration we will use Rank Selection method for better exposure to the readers.

With a threshold value of 50%, three best chromosomes shall be selected in the mating pool while the remaining chromosomes shall be discarded. Thus the new population after applying threshold will be as given in Table 3.10.

Table 3.10
Objective Function Values of Initial Chromosomes

Population	Actual Values	Obj. Fn. Value
Chromosome 1	600 200 −200 −360 −120	147.4149
Chromosome 2	120 −40 −600 −120 360	131.3716
Chromosome 3	120 40 120 −520 40	76.2643

So now we have the three best chromosomes from which we need to select the parents to generate three child chromosomes. This means that a pair of parents will generate a single child and not two as discussed in the earlier sections. This is because of small population and the number of child chromosomes being generated are odd, i.e., 3. From here we can also observe that GA is quite flexible and the parameters can be varied as per the requirements.

To create a Roulette Wheel, we need to find probabilities of each chromosome. Roulette Wheel probability distribution cannot be based on minimization conditions as it will allocate minimum probability range to the best chromosome. Therefore we convert the objective function values into a maximization function. The method used is:

New Objective Fn Value = $\frac{1}{(1+Obj.Fn.Value)}$

The objective function values of the three chromosomes is thus found to be: 0.0067, 0.0076 and 0.0129.

Now using Eq 3.1, the probabilities of each chromosome is calculated.

For the first chromosome, $p(1) = \frac{0.0067}{(0.0067+0.0076+0.0129)} = 0.2474$

Similarly, $p(2) = \frac{0.0076}{(0.0067+0.0076+0.0129)} = 0.2774$

And $p(3) = \frac{0.0129}{(0.0067+0.0076+0.0129)} = 0.4752$

The ranges of these chromosomes shall be p(1) from 0 to 0.2474, for p(2) it will be from 0.2475 to 0.5248 (as 0.2474 + 0.2774 = 0.5248) and for p(3) it will be from 0.5249 to 1.

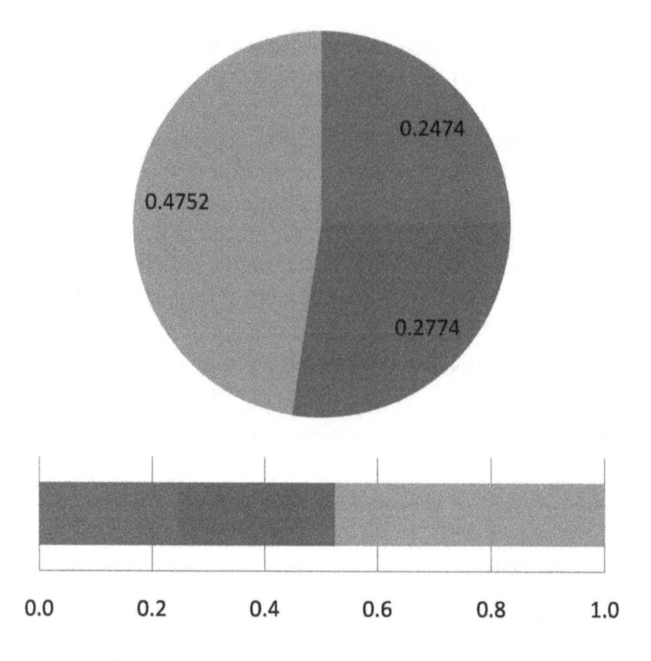

Figure 3.23: Roulette Wheel for Iteration 1

The Roulette Wheel for the above distribution can be represented as in Fig 3.23:

Random values are generated to select parents for child chromosome reproduction. These values are generated in pairs to select two parent chromosomes for generating one child.

The random values are represented in Table 3.11 along with the chromosomes that they represent. Three pairs of parent chromosomes are selected to produce three child chromosomes.

Table 3.11
Parent Selection for Iteration 1

First Pair		Second Pair		Third Pair	
Random No.	Chromosome No.	Random No.	Chromosome No.	Random No.	Chromosome No.
0.5804	3	0.1551	1	0.1075	1
0.3381	2	0.3470	2	0.7649	3

The random numbers generated for parent selection can be plotted to the chromosome number through the Fig 3.23 and also from the probability ranges mentioned before. From Table 3.11, it can be observed that the first child will be formed from chromosomes 3 & 2, the second from chromosomes 1 & 2 and last from chromosomes 1 & 3. One point to be noted here is that, the initial random number generated for the second parent in second pair was 0.0580, which would have led both the parents in the second pair to be the same, thus another random number was generated to get a different parent. It should be thus kept in mind to not repeat parent chromosomes for a crossover, other wise the whole exercise would go waste. Also in this example all the parent chromosomes are getting a chance to reproduce, since the size of the example is small. For optimization problems involving more number of chromosomes, some of the parent chromosomes may not get a chance to reproduce.

Now that the parent chromosomes have been decided we move towards the next part of crossover. The crossover point is randomly selected. Random numbers generated are 8, 18 and 13 for the three crossovers. The three crossovers are shown in Fig 3.24.

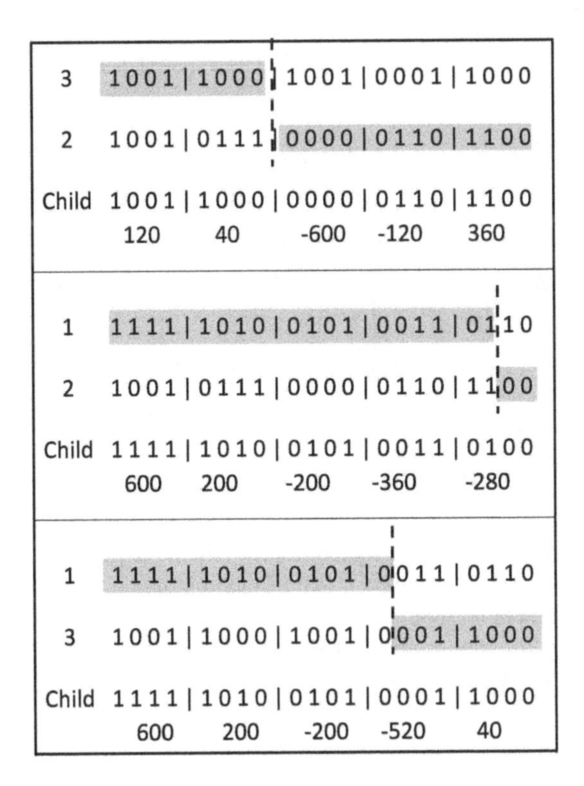

Figure 3.24: Crossover for Iteration 1

In the figure, all the crossovers are clearly visible at the stated positions. The name of the parent pairs is given on the left side above each child. The decoded values of these child chromosomes are taken from Table 3.8 and presented below each child chromosome, respectively. The three child chromosomes are thus produced.

The last process in the first iteration would be mutation. In this example, a two point mutation is performed for each child chromosome. The positions for mutation is generated randomly. The random numbers generated for the three child chromosomes are: (14,15), (4,17) and (2,6), respectively.

The mutated child chromosomes along with their decoded values are shown in Fig 3.25. The mutated values are shown through the arrows in the figure.

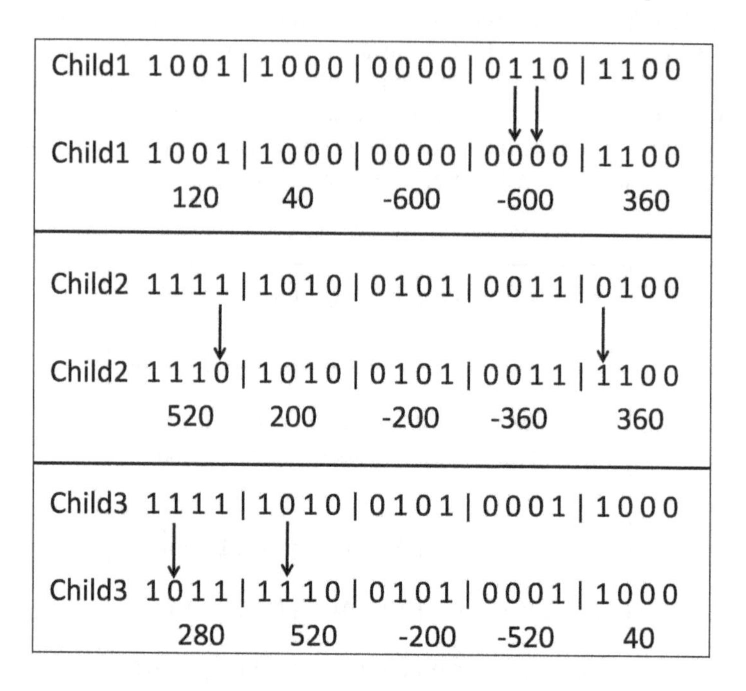

Figure 3.25: Mutation for Iteration 1

The new population in GA would consists of the three parents and the three child chromosomes, thus taking the total count of the population to 6, which was also the size of the original population.

The objective function values for the new chromosomes have to be evaluated. The new population along with their objective function values are presented in Table 3.12.

Table 3.12
Objective Function Values of Population after Iteration 1

Population	Actual Values	Obj. Fn. Value
Chromosome 1	600 200 −200 −360 −120	147.4149
Chromosome 2	120 −40 −600 −120 360	131.3716
Chromosome 3	120 40 120 −520 40	76.2643
Chromosome 4	120 40 −600 −600 360	217.4086
Chromosome 5	520 200 −200 −360 360	153.3796
Chromosome 6	280 520 −200 −520 40	165.9253

As can be observed, the new chromosomes do not have better objective function values. The parent chromosomes are still better than the new child chromosomes.

Let us start the next iteration to see, if some better results can be obtained. The first process is the selection of parents. As in the previous iteration, truncation selection with 50% threshold is used, with the Roulette Wheel method being replaced by Rank Selection method.

The new set of chromosomes selected for the mating pool are the same as the previous iteration, as the new child chromosomes do not have better objective function values than all the parent chromosomes and are given in Table 3.10.

Since the Rank Selection method is being used, the chromosomes are ranked according to their objective function values with the best chromosome getting the highest rank and the worst chromosome getting rank 1. The chromosome 3 gets rank 3 with an objective function value of 76.2643, chromosome 2 gets the rank 2 with objective function value of 131.3716 and first chromosome gets rank 1 with objective function value of 147.4149. The probabilities are distributed based on Eq 3.2.

Thus $p(1) = \frac{1}{3*2} = 0.1667$

And $p(2) = \frac{2}{3*2} = 0.3333$

Lastly $p(3) = \frac{3}{3*2} = 0.5$

Let's view the above probability distribution through a pie chart and a bar chart, through Fig 3.26.

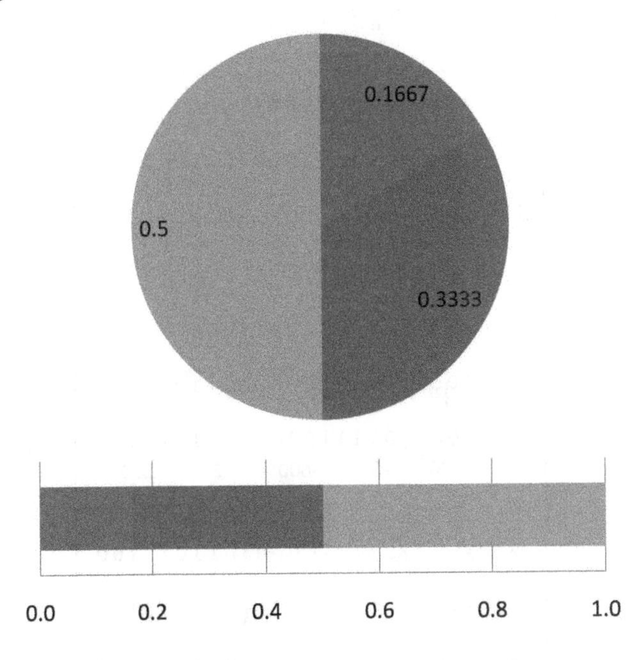

Figure 3.26: Rank Selection for Iteration 2

It can be observed that the probabilities of first chromosome/worst chromosome changes a lot as compared to the Roulette Wheel Selection method. The probability ranges of the chromosomes will be – (0 to 0.1667) for chromosome 1, (0.1668 – 0.5) for chromosome 2 and (0.5 – 1) for chromosome 3.

Random numbers are generated to select pairs of parent chromosomes required to reproduce the child chromosomes. The random numbers and the parent pairs are tabulated in Table 3.13.

Table 3.13
Parent Selection for Iteration 2

First Pair		Second Pair		Third Pair	
Random No.	**Chromosome No.**	**Random No.**	**Chromosome No.**	**Random No.**	**Chromosome No.**
0.9902	3	0.716	3	0.1777	2
0.2878	2	0.2956	2	0.1033	1

The first and second pairs of the parents are the same, but then the crossover point and mutation will make different child chromosomes.

Three random numbers are generated for the three crossover points applied on three pairs of parent chromosomes. These random positions are: 15, 2 and 12. The crossover procedure is shown in Fig 3.27.

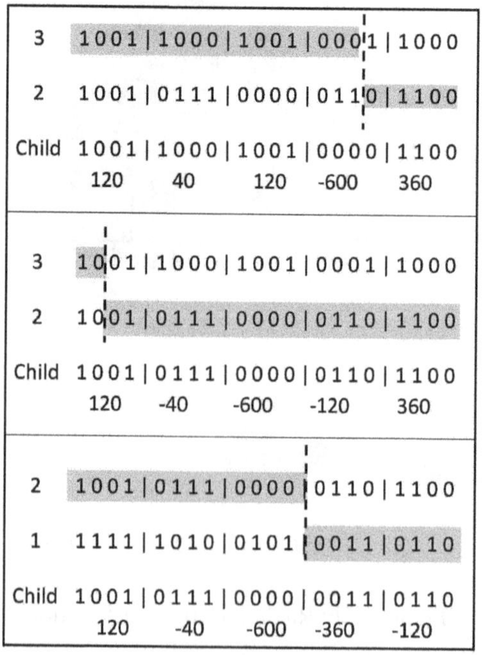

3 1001 | 1000 | 1001 | 0001 | 1000
2 1001 | 0111 | 0000 | 0110 | 1100
Child 1001 | 1000 | 1001 | 0000 | 1100
120 40 120 -600 360

3 1001 | 1000 | 1001 | 0001 | 1000
2 1001 | 0111 | 0000 | 0110 | 1100
Child 1001 | 0111 | 0000 | 0110 | 1100
120 -40 -600 -120 360

2 1001 | 0111 | 0000 | 0110 | 1100
1 1111 | 1010 | 0101 | 0011 | 0110
Child 1001 | 0111 | 0000 | 0011 | 0110
120 -40 -600 -360 -120

Figure 3.27: Crossover for Iteration 2

The last process of the iteration is Mutation. As in the previous iteration, a two point mutation shall be performed. The random positions for these mutations are: (3, 17) for child 1, (6,17) for child 2 and (5, 18) for child 3. The mutation is presented in Fig 3.28.

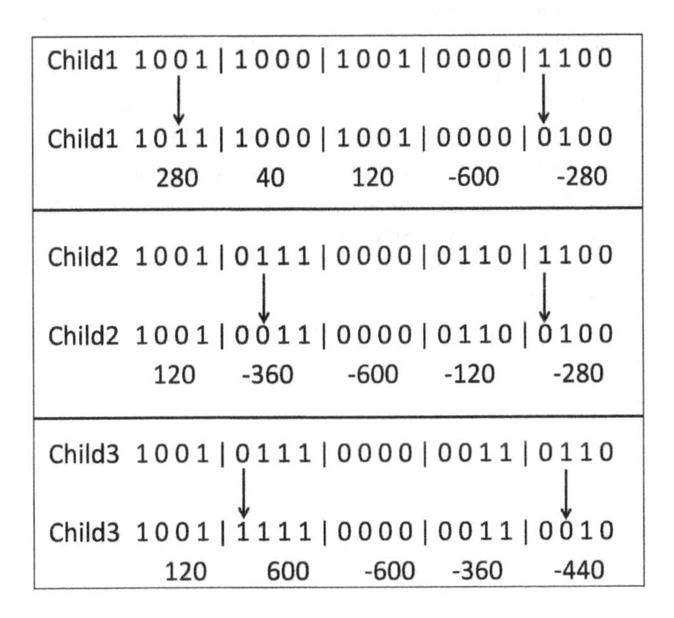

Figure 3.28: Mutation for Iteration 2

The decoded child chromosomes are also represented in Fig 3.28. It can be observed that even though the first two child chromosomes were produced from the same pair of parents, they are quite different in content from each other. Thus a new population has now been created with the old parent chromosomes and new child chromosomes. These chromosomes are tabulated in Table 3.14 along with their objective function values.

Table 3.14
Objective Function Values of Population after Iteration 2

Population	Actual Values	Obj. Fn. Value
Chromosome 1	600 200 −200 −360 −120	147.4149
Chromosome 2	120 −40 −600 −120 360	131.3716
Chromosome 3	120 40 120 −520 40	76.2643
Chromosome 4	280 40 120 −600 −280	134.2181
Chromosome 5	120 −360 −600 −120 −280	149.7318
Chromosome 6	120 600 −600 −360 −440	265.5332

Even though no major improvement has been made, but one of the child chromosome, i.e., the 4th chromosome, is now part of the best three with an objective function value of 134.2181.

The iterations are repeated until the stopping criterion has been met. Each iteration shall have the selection, crossover and mutation phases. In this example, because of lack of variety and small population, we were unable to achieve major improvements; however, with increase in population the improvement would be clearly visible. Also, since the step size is too large, the output is not reducing faster.

GA was executed on a similar basis for 200 iterations while taking only Rank Selection method. The variation in output for the best chromosome is presented in Fig 3.29. The best chromosome in the population approximately reaches the best possible solution with the given step size (i.e., the best possible solution is when all input variables have a value of 0, however, 0 is not a possible value for input variable). The minimum objective function value comes out to be 6.04631 with the chromosome being given as, 40 −40 120 −40 −40.

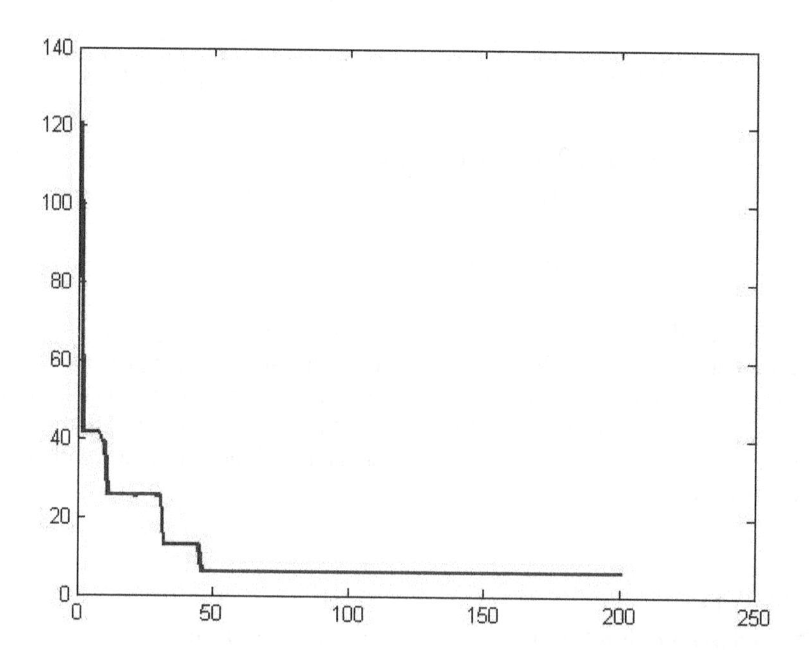

Figure 3.29: Iteration wise variation of Best Chromosome

3.7 VARIANTS AND HYBRID

This section presents some of the variants and hybrid versions of GA. By variant it is assumed that some changes are proposed in the search process of GA itself, whereas in hybrid, GA is combined with some other optimization technique. There may be multiple references for one variant, however, we will stick to a single reference to understand the concept.

3.7.1 VARIANTS

Some of the variants of GA are enlisted and explained below:

1. **Real Coded GA**: The real coded GA is the most important variant of GA. In conventional GA, the chromosomes are represented in binary (0,1) and this method of representation is very rigid. The binary coded GA is not feasible for many of the optimization problems as they contain continuous variables. Also the coding and decoding of chromosomes takes lot of computational time. In real coded GA, the numbers are directly placed instead of binary values. The various references using real coded GA are: [15, 25, 32, 105].
2. **Crossover-Based Variants**: A comparative study of different crossover methods for real coded GA is presented in [39]. Some of the methods used for crossover in real coded GA are listed below:

   a. **Linear Crossover**: In [171], Linear Crossover was initially introduced. The method is an extension of arithmatic crossover method. In Linear crossover, two parents are used to generate three child chromosomes, out of which the two best child chromosomes are selected. The child chromosome are generated using the formulae: $(0.5x_1 + 0.5x_2)$, $(1.5x_1 - 0.5x_2)$ and $(-0.5x_1 + 1.5x_2)$, where x_1 and x_2 are individual input variables of parents 1 and 2, respectively. An example of this crossover techniques is presented in Fig 3.30. The limits for individual values is assumed to be from (-10) to (10). It can be observed that the child chromosomes have values crossing the allowable limits. In such cases, the values of the input variables can be set to their maximum or minimum limits, respectively. In some references, it has been suggested that when input variable values cross the limits then their values should be taken from child 1.

Parent 1	4	4	-10	-9	10	6	8	-10	-7	9
Parent 2	-8	9	7	5	1	-1	2	6	-6	1

Child 1	-2	6.5	-1.5	-2	5.5	2.5	5	-2	-6.5	5	$0.5x_1+0.5x_2$
Child 2	10	1.5	-18.5	-16	14.5	9.5	11	-18	-7.5	13	$1.5x_1-0.5x_2$
Child 3	-14	11.5	15.5	12	-3.5	-4.5	-1	14	-5.5	-3	$-0.5*x_1+1.5*x_2$

Figure 3.30: Linear Crossover

   b. **Blend Crossover (BLX - α)**: Blend Crossover is presented in [40] and is an extended version of Linear crossover. The relation between Linear Crossover and the Blend Crossover is shown in Fig 3.31. From the figure and above explanation of Linear Crossover, it can be observed that the

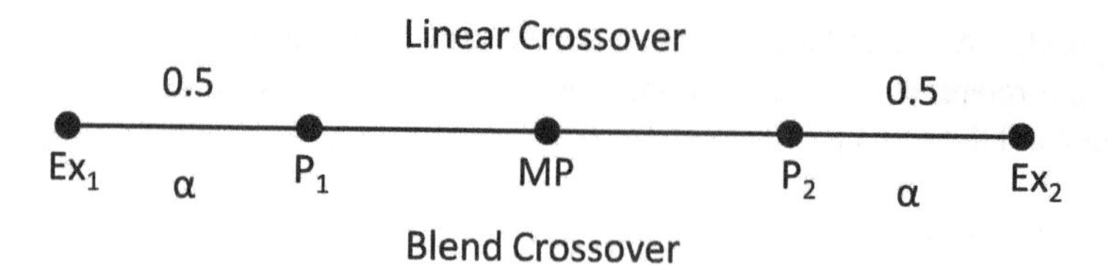

Figure 3.31: Blend Crossover

child chromosomes in Linear Crossover lie on the line joining parents P_1 and P_2. The three child chromosomes generated by Linear Crossover are present at positions MP (i.e., mid point position in space between the parents), Ex_1 (i.e., a position achieved by extending the line joining the parents by a factor of 0.5 on the side of P_1) and Ex_2 (i.e., a position achieved by extending the line joining the parents by a factor of 0.5 on the side of P_2). Blend Crossover works on the same principle except that 0.5 is replaced by α. The value of α can be randomly varied, but it is proposed to keep this value in between 0 and 0.5.

c. **Simulated Binary Crossover (SBX)**: The SBX can be referenced for its working through [29, 31]. This crossover method is an extension of the earlier introduced crossover methods but concentrates on proper distribution of child chromosomes. The equations used in SBX are Eqs 3.6, 3.7 and 3.8.

$$\beta_q = \begin{cases} (2u)^{\frac{1}{\eta+1}} & if \quad u \leq 0.5; \\ \left(\frac{1}{2(1-u)}\right)^{\frac{1}{\eta+1}} & otherwise; \end{cases} \tag{3.6}$$

$$y_1 = 0.5[(1+\beta_q)x_1 + (1-\beta_q)x_2)] \tag{3.7}$$

$$y_2 = 0.5[(1-\beta_q)x_1 + (1+\beta_q)x_2)] \tag{3.8}$$

where u is a random number between 0 and 1 and η is a non negative real number. A large value of η will create child chromosomes near to the parent chromosomes and a smaller value of η will create distant solutions. The process of creating child chromosomes requires a random number generation for 'u' in between 0 and 1. Next decide the value of η based on the exploration or exploitation to be performed. After getting these values, β_q is found using Eq 3.6. Lastly, individual child chromosome input values are found using Eqs 3.7 and 3.8.

An example of SBX calculation is presented through Fig 3.32. The figure shows the child chromosome evaluation from the same parent chromosomes while using different values of 'u'. In both the cases, the value of η is taken as 2. The process followed in evaluating child chromosomes are explained through the following calculations:

Case 1:
Generate random value of u - 0.16
Assume η = 2
$\beta = (2*u)^{\frac{1}{\eta+1}} = (2*0.16)^{\frac{1}{(2+1)}} = 0.32^{0.33} = 0.684$
For child 1, first input value:
$y_1 = 0.5[(1+\beta_q)x_1 + (1-\beta_q)x_2)] = 0.5[(1+0.684)*6+(1-0.684)*0] = 5.052$
For child 2, first input value:
$y_2 = 0.5[(1-\beta_q)x_1 + (1+\beta_q)x_2)] = 0.5[(1-0.684)*6+(1+0.684)*0] = 0.948$
Similarly, values of other input variables can be evaluated.
Case 2:
Generate random value of u - 0.78
Assume η = 2
$\beta = \left(\frac{1}{2(1-u)}\right)^{\frac{1}{\eta+1}} = \left(\frac{1}{2(1-0.78)}\right)^{\frac{1}{2+1}} = 2.2727^{0.33} = 1.3148$
For child 1, first input value:
$y_1 = 0.5[(1+\beta_q)x_1 + (1-\beta_q)x_2)] = 0.5[(1+1.3148)*6+(1-1.3148)*0] = 6.944$
For child 2, first input value:
$y_2 = 0.5[(1-\beta_q)x_1 + (1+\beta_q)x_2)] = 0.5[(1-1.3148)*6+(1+1.3148)*0] = -0.94$

Parent 1	6	-3	0	1	9	10	7	-8	2	2
Parent 2	0	2	1	4	5	-8	-3	-10	1	7

u 0.16
η 2
β 0.684

Child 1	5.052	-2.21	0.158	1.474	8.368	7.156	5.42	-8.32	1.842	2.79
Child 2	0.948	1.21	0.842	3.526	5.632	-5.16	-1.42	-9.68	1.158	6.21

u 0.78
η 2
β 1.3148

Child 1	6.944	-3.79	-0.16	0.528	9.63	12.83	8.574	-7.69	2.157	1.213
Child 2	-0.94	2.787	1.157	4.472	4.37	-10.8	-4.57	-10.3	0.843	7.787

Figure 3.32: Simulated Binary Crossover

The readers are encouraged to try different combinations of 'u' and η.

d. **Simplex Crossover (SPX)**: SPX is an extension of the Blend Crossover [161]. The benefit associated with SPX is that it can be applied to a crossover involving more than or equal to two parents. As in the case of Blend Crossover, the extension is given by 'α', here it is represented by 'ε'. However, one needs to find the center of mass of the number of chromosomes involved. A three parent SPX can generate three child chromosomes and so on. The SPX applied for a three parent crossover is shown in Fig 3.33. The method involves finding the centroid of the

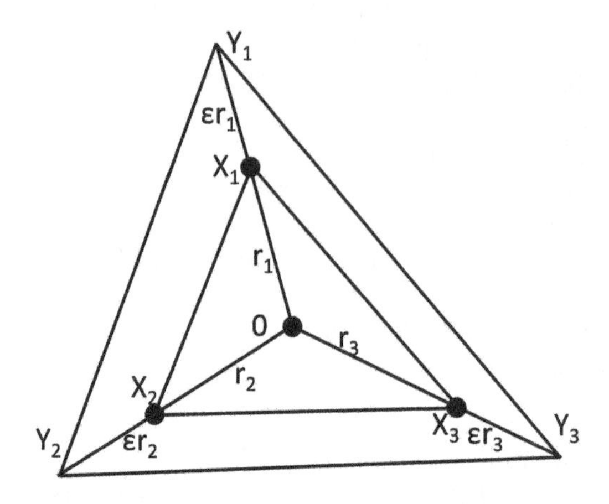

Figure 3.33: Simplex Crossover

shape formed by the number of parents. Then the child chromosomes are generated on the line joining the centroid and the parent. In the Fig 3.33, parent chromosomes are represented by X_1, X_2 and X_3 while 'O' gives the position of the centroid. The lines joining the centroid and the parent chromosomes are given by r_1, r_2 and r_3, respectively. These lines are extended by a factor 'ε' (similar to 'α' in Blend Crossover) where $\varepsilon \geq 0$. The child chromosomes can be generated on these lines given by Y_1, Y_2 and Y_3, respectively. Thus for a three parent SPX Crossover, three child chromosomes are obtained.

e. **Parent Centric Crossover (PCX)**: A parent centric chromosome generation is applied in PCX [30]. First a mean vector ($\vec{g}$) is evaluated for the parents (μ) chosen for reproduction. For each child chromosome, one parent ($\vec{x}^p$) is chosen. The allocation of parent to the child chromosome is done with equal probability. Next the direction vector ($\vec{d}^p$) is obtained through the difference between the parent chromosome and the mean. Then, from the remaining parents, i.e., parents other than the chosen parent, perpendicular distances (D_i) to line $\vec{d}^p$ are found and the average ($\vec{D}$) is obtained. The equation used to create child chromosomes is Eq 3.9.

$$\vec{y} = \vec{x}^p + w_\zeta \vec{d}^p + \sum_{i=1, i \neq p}^{\mu} w_\eta \vec{D} \vec{e}^i \tag{3.9}$$

In Eq 3.9, w_ζ and w_η are zero-mean normally distributed variables with variance of σ_ζ^2 and σ_η^2, respectively. Also $\vec{e}^i$ are orthonormal bases through the space and normal to $\vec{d}^p$.

The PCX method creates child chromosomes near the parents. The method is too lenghty in terms of calculations involved.

f. **Triangular Crossover (TC)**: In Triangular Crossover [38], three parents are selected for crossover given by X_1, X_2 and X_3, respectively. Next, three random numbers are chosen in between 0 and 1. They are represented as r_1, r_2 and r_3. These random numbers should be such that $r_1 + r_2 + r_3 = 1$. The child chromosomes in TC are evaluated as given by Eq 3.10:

$$Y_1 = (r1 \times X_1) + (r2 \times X_3) + (r3 \times X_2) \tag{3.10a}$$

$$Y_2 = (r1 \times X_2) + (r2 \times X_1) + (r3 \times X_3) \tag{3.10b}$$

$$Y_3 = (r1 \times X_3) + (r2 \times X_2) + (r3 \times X_1) \tag{3.10c}$$

It is suggested that while selecting parents for TC, two should be feasible and one should be infeasible. This creates higher probability to obtain feasible child chromosomes. In this method, the child chromosomes are generated nearer to the boundary of search space than their parents.

There are multiple other crossover methods like heuristic crossover, unimodal normal distribution crossover (UNDX), etc.

3. **Selection-Based Variants**: In [62], a number of variants have been presented based on the creation of new chromosome population. The methods are explained below:

   a. **Generational GA (GGA)**: In this process the whole population is replaced by the new child population. No elitism is considered in this case. Two parents are selected through Tournament selection to form a single child chromosome. The process is repeated until the child population is equal to the existing population.
   b. **Steady State GA (SSGA)**: In SSGA, two parents are used to create a new child which will replace the worst chromosome in the existing population.
   c. **Steady Generational GA (SGGA)**: SGGA is similar to SSGA, except that in SGGA, a random solution is replaced by the child chromosome instead of the worst chromosome in the population.
   d. **$(\mu + \mu)$ - GA**: In this method of creating a new generation, the child chromosomes are generated similar to SGA until the number of these child chromosomes is equal to the existing population. The population for the next generation is formed by selecting the best chromosomes from the existing population and the child population.

4. **Quantum GA (QGA)**: Quantum theory is based on the complicated behavior of the atomic structure. Quantum computer works on the concept of quantum bits (qubits). Qubits may be either zero or one or superposition of both. This concept comes from the fundamentals of "quirks" of quantum mechanics. In QGA, "the concept of superposition, entanglements and intervention are taken from quantum computing while selection, crossover and mutation are taken

from GA" [72]. A single chromosome is given in Eq 3.11:

$$p_j(t) = \begin{bmatrix} \alpha_1(t) & \alpha_2(t) & \dots & \alpha_m(t) \\ \beta_1(t) & \beta_2(t) & \dots & \beta_m(t) \end{bmatrix} \tag{3.11}$$

where a Q bit solution is represented by an 'm' Q bit string, such that $|\alpha_i|^2 + |\beta_i|^2 = 1$ and $i \in [1,m]$. Once the chromosome have been properly represented, two operators are applied in QGA. The Quantum rotation gate is defined as in Eq 3.12.

$$U(\Delta\theta) = \begin{bmatrix} \cos(\Delta\theta) & -\sin(\Delta\theta) \\ \sin(\Delta\theta) & \cos(\Delta\theta) \end{bmatrix} \tag{3.12}$$

where $\Delta\theta$, is the rotation angle, and its magnitude as well as direction are very crucial for the efficient execution of the optimization technique. The chromosome shall be rotated using quantum rotation operation for each Q bit such that it moves towards the optimal chromosome. This rotation is denoted by Eq 3.13:

$$\begin{bmatrix} \alpha(t+1) \\ \beta(t+1) \end{bmatrix} = U(\Delta\theta) \begin{bmatrix} \alpha(t) \\ \beta(t) \end{bmatrix} \tag{3.13}$$

Next, Quantum mutation operator is applied, to improve the diversity of the population. The mutation process is represented through Eq 3.14.

$$\begin{bmatrix} \alpha'(t) \\ \beta'(t) \end{bmatrix} = \begin{bmatrix} \beta(t) \\ \alpha(t) \end{bmatrix} \tag{3.14}$$

5. **Competitive Co-Evolutionary QGA (CCQGA)**: In [51], a Competituve Co-evolutionary Quantum GA (CCQGA) is implemented through three different methods. In this research work, QGA is used along with different scenarios. It is assumed that the size of the population can keep varying.
   The first method, involves the competitive hunter methodology, in which the spotted Owls compete with the Hawks. The populations of both the species increase and decrease based on their competition and interactions with each other, since their source of food are the same. The competitive fitness of each individual plays an important role in its survival.
   In the next method proposed, cooperative surviving is discussed between White ants and Flagellates. In this scenario, the population of one species can thrive if they cooperate with the other. The more a species cooperates with the other, the more are the chances of its population getting increased.
   In the last scenario, big fish eating small fish is considered with the big fish as the whales and the small fish as the fingerlings. The whales eat the fingerlings and thus the population size of fingerlings also include a death rate factor.
6. **Opposition-Based GA**: Opposition-based variants survive on the concept of bringing exploration into the search process by generating solutions which are on the opposite side of the search space for the chromosomes having better objective function values. "An opposite position in the search space can be defined as a point where all the dimensions of this point are replaced by their

respective opposites" [59]. The opposite of a given chromosome is evaluated as in Eq 3.15:

$$x_i^{'} = a_i + b_i - x_i \tag{3.15}$$

where $x_i^{'}$ is the opposite chromosome, a_i and b_i are the limits of the particular dimension, i.e., input variable and x_i is the current position of a chromosome.

7. **Parallel GA (PGA)**: In PGA, different search trajectories are employed to perform the search process in parallel [100]. A trajectory can be considered as a sub population. After every 'k' specified iterations the best solution from a subpopulation is sent to its neighboring subpopulation if it is not progressing for a specific number of iterations.
8. **Hierarchical GA (HGA)**: The HGA was developed on the basis of PGA, however, the subpopulations are arranged in a hierarchical topology [138]. The benefit of HGA is that the better solutions move upwards in the heirarchy. The top layer makes very small movements and very less exploration is present. On the other side, the lower most layer concentrates more towards exploration of the search space.
9. **Structured GA (sGA)**: In sGA, the data in the chromosome are stored in a hierarchical manner, termed as a genome structure [26]. A simple two-level structure is presented in Fig 3.34. The chromosome is therefore represented as $A = < S_1, S_2 >$ where the length of S_2 is an integer multiple of S_1. In this variant, the genes may be active or passive. The high-level genes are able to activate or deactivate the lower level genes. A change in a gene at a higher level would affect multiple changes at the lower level of hierarchy. The passive genes are not removed from the population.

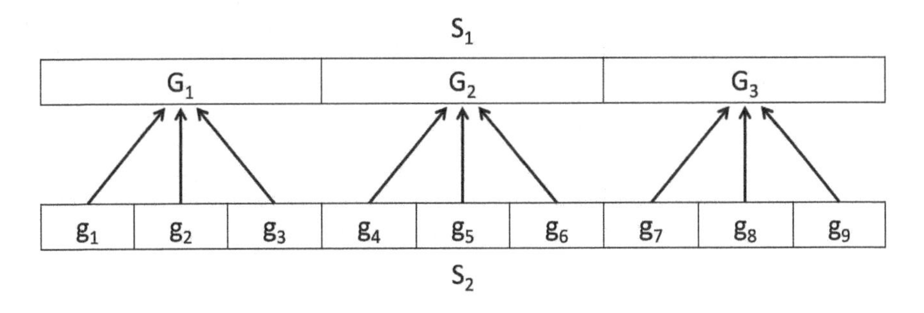

A = <S1,S2> = [(G1, G2, G3), (g1, g2, g3, g4, g5, g6, g7, g8, g9)]

Figure 3.34: Chromosome Representation in Structured Genetic Algorithm

10. **Adaptive GA (AGA)**: An AGA version of GA is presented in [145]. It is proposed to vary the probabilities of the crossover p_c and mutation p_m. The chromosomes having better objective function values are not disturbed, however, the other chromosomes are disrupted. The proposed ranges of p_c and p_m are 0.5 to 1 and 0.005 to 0.05, respectively. The expressions for these probabilities

are given in Eq 3.16.

$$p_c = k_1 \left(\frac{f_{max} - f'}{f_{max} - \vec{f}} \right), \qquad f' \geq \vec{f} \tag{3.16a}$$

$$p_c = k_3, \qquad f' < \vec{f} \tag{3.16b}$$

$$p_m = k_2 \left(\frac{f_{max} - f}{f_{max} - \vec{f}} \right), \qquad f \geq \vec{f} \tag{3.16c}$$

$$p_m = k_4 \qquad f < \vec{f} \tag{3.16d}$$

Such that $k_1, k_2, k_3, k_4 \leq 1.0$ and where f_{max} is the best objective function value, $\vec{f}$ is average objective function value, f is the objective function value of the chromosome and f' is the better objective function value that needs to be crossed.

11. **Messy GA (mGA)**: The mGA [48] has a unique property of dealing with variable length chromosome strings. A summation of subfunctions is used to find the overall function value. The crossover operator is replaced with two new operators termed splice and cut. The mutation performed is termed allelic mutation. Fig 3.35 shows the working of Cut and Splice operators. The cut op-

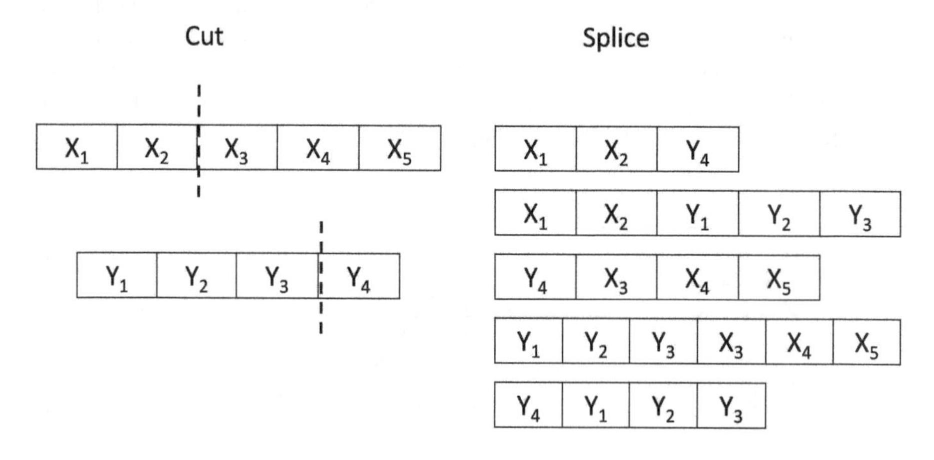

Figure 3.35: Cut and Splice Operator in Messy Genetic Algorithm

eration is shown in the left side of the figure. The position where the cut is to be placed can be decided based on various factors like grouping of input variables, etc. The right side of the figure shows the splice operation, where some of the possible outputs/combination are shown and can be further expanded.

12. **Clustered GA (CGA)**: CGA is presented in [186]. The proposed method uses K-means algorithm to create clusters in the population and also introduces fuzzy system to adjust the probabilities for crossover and mutation.

13. **Multiobjective GA (MOGA)**: A multiobjective optimization variant of GA is introduced in [101]. The parents selected for a crossover process are based on weighted sums of different objective functions. The weights attached to these objective functions are randomly specified for each selection. It also implements elitism.

3.7.2 HYBRID

A hybrid optimization considers the better characteristics of two or more optimization techniques and merges these techniques. A combination of such hybrid optimization techniques involving GA is presented in Table 3.15. There are multiple hybrid versions of GA, out of which some are listed below.

Table 3.15
Hybrid GA

Sr. No.	Hybrid With	Reference
1	Particle Swarm Algorithm	[45, 121, 140, 182]
2	Differential Evolution	[54, 82]
3	Ant Colony Optimization	[42, 89]
4	Chaotic	[183]
5	Artificial Bee Colony	[120]
6	Shuffled Frog Leaping Algorithm	[63, 66]
7	Simulated Annealing	[135]
8	Tabu Search	[173, 181]
9	Gravitational Search Algorithm	[46, 69]
10	Sequential Quadratic Programming	[75, 91, 170, 179]
11	Immune Algorithm	[14]
12	Bacteria Foraging	[37]
13	Taguchi	[130, 159, 160]

4 Differential Evolution

4.1 INTRODUCTION

Differential Evolution (DE) was introduced by Storn and Price [146] in 1995 and a detailed study was presented by them again in [147]. DE is marked to efficiently handle continuous search spaces. DE has been researched widely over a period of years, however, due to the presence of Genetic Algorithm (GA) and Particle Swarm Optimization (PSO), it has been somewhat undermined.

DE is a simple optimization technique and requires little efforts on the side of the programmer to implement this method. DE has some very simple drawbacks as observed from the literature, like:

- There are too many random factors involved and thus the search is directionless.
- The value of 'F' in Eq 4.1 could have been defined properly (similar to ω in PSO).
- The value of 'CR' is not properly defined in Eq 4.2.
- The number of vectors generated should be at least '4'.
- The random vectors selected in Eq 4.1 are totally random and some method could have been used to nominate vectors for this equation for better performance.

These drawbacks may have been dealt with by other references, however, this chapter shall be discussed and presented based on [147].

4.2 TERMINOLOGY

The terminology used in DE is presented as a list in this section. The terminology of DE is not easy to follow but their analogous terms can be found in other optimization techniques.

1. **Vector**: A potential solution in DE is termed as a vector. It will have representation of all the input variables. It is similar to a chromosome in GA or a particle in PSO. It is denoted by 'x'.
2. **NP**: It is a collection of vectors. It basically represents a population.
3. **G**: The iterations in DE are termed as generations and are represented by 'G'.
4. **Target Vector**: The vector which is under consideration is termed as the Target Vector. It is denoted as '$x_{i,G}$' where 'i' is vector number and 'G' is generation count.
5. **Mutant Vector**: A randomly generated vector based on three different vectors is a mutant vector. The mutant vector is represented by '$v_{i,G}$'.
6. **Mutation**: Process of generation of a random vector, i.e., a mutant vector is termed as Mutation. It is not similar to the mutation process in GA.

DOI: 10.1201/b22647-4

7. **Trial Vector**: A vector created by the combination of the existing vector and its mutant vector is termed as Trial Vector. It is denoted as '$u_{i,G}$'.
8. **Crossover**: Process of generation of a trial vector is termed as Crossover.
9. **Selection**: Process of selecting the target vector or trial vector for the next generation is termed as Selection.

4.3 FUNDAMENTAL CONCEPT

DE does not seem to follow any natural process. The method involved in solution is similar to many other EOAs, involving mutation, crossover and selection processes. DE is a very simple optimization method with less number of equations to be executed and therefore less complexity. However, unlike PSO the amount of randomness in the method is too much.

DE, similar to GA, revolves around three processes of mutation, crossover and selection. However, these processes are not implemented as it is from GA, but in principle. In DE, individual potential solutions are termed as vectors. A random set of vectors is initially generated. In DE, it is by default assumed that these vectors are spread throughout the search space.

Mutation

After the generation of vectors, randomly, we enter the iteration process. In the iterations, Mutation is the first process. In mutation, a mutant vector is generated for each vector in 'NP' using Eq 4.1.

$$v_{i,G+1} = x_{r_1,G} + F \cdot \left(x_{r_2,G} - x_{r_3,G}\right) \tag{4.1}$$

where '$v_{i,G+1}$' is the ith mutant vector corresponding to i^{th} target vector for next iteration. 'r_1', 'r_2' and 'r_3' are random integers which are less than NP and mutually different. Also these random integers should not be the same as 'i'. 'F' is a real and constant factor such that $0 < F < 2$.

Crossover

In the next step, Crossover is performed. In crossover, a trial vector 'u' is formed with the help of target vector, i.e., x_i and the mutant vector 'v' generated through Eq 4.1. For crossover to take place and to generate a trial vector, we need to loop through all the input variables. The trial vector shall be populated either by the target vector value or the mutant vector value for a particular input variable depending in Eq 4.2.

$$u_{ji,G+1} = \begin{cases} v_{ji,G+1} & \text{if } (\text{randb}(j) \le CR\,) \text{ or } j = \text{rnbr}(i) \\ x_{ji,G} & \text{if } (\text{ randb } (j) > CR) \text{ and } j \ne \text{ rnbr}(i) \end{cases} \tag{4.2}$$
$$j = 1, 2, \ldots, D$$

where 'randb' is a random number which lies between 0 and 1 and is generated for every input variable, given by 'j'. CR represents the crossover constant, which also lies between 0 and 1. 'rnbr(i)' represents a random integer value between 0 and 'D' (total number of input variables). This term is used to have at least one term of mutant vector in the trial vector. After generation of the trial vector, the input variable limits are checked and its objective function value is obtained.

Selection

Selection is the last process in a DE iteration. In the selection process, one out of trial vector and target vector is chosen based on the greedy selection criterion. If the objective function value of trial vector is better than the target vector, then the target vector is replaced by the trial vector, otherwise the target vector is retained.

This process is continued until the stopping criterion is fulfilled.

4.4 ALGORITHM AND PSEUDOCODE

A detailed algorithm of DE is presented in Algorithm 4.1. In DE, the first step deals with generation of random vectors. We need a nested loop structure with the outer one repeating for maximum number of iterations and the inner one repeating through 'NP'.

Algorithm 4.1 Differential Evolution

```
START
Initialize NP random vectors
for 1 to Maximum Number of Iterations do
   for 1 to NP do
      Generate Mutant Vector using Eq 4.1
      for 1 to D do
         if randb(j) ≤ CR or j==rnbr(i) then
            u_{ji,G+1} = v_{ji,G+1}
         else
            u_{ji,G+1} = x_{ji,G}
         end if
         Find Objective Function Value for Trial Vector 'u'
         if f(u) < f(x) (for minimization problem) then
            x = u
         end if
      end for
   end for
end for
Display Result
STOP
```

In each iteration, three steps of mutation, crossover and selection are involved. First the mutant vector is generated through Eq 4.1 for each target vector present in NP. Next the crossover process is applied between the mutant vector and its respective target vector using Eq 4.2. The trial vector is generated through the crossover process, represented by 'u'. Lastly to apply the selection process, decision control statement 'if' is used. The objective function values of target vector (f(x)) and trial vector (f(u)) are compared, for selecting the better vector through greedy criterion.

The stopping criterion is checked. The iterations are repeated until the stopping criterion is met. At the end of the optimization process, the results are displayed.

4.5 FLOWCHART

The flowchart of DE is shown in Fig 4.1. From the flowchart it can be seen that in DE, we first generate random vectors which are spread throughout the search space. The iteration process of DE is classified into three parts: Mutation, Crossover and Selection. The iterations shall continue until the stopping criterion is satisifed. It can

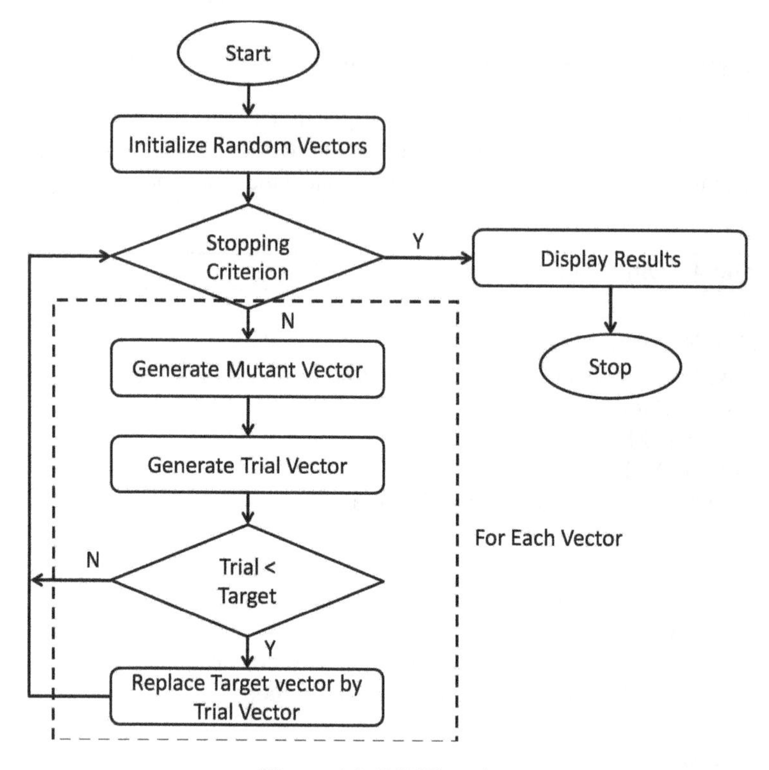

Figure 4.1: DE Flowchart

be observed from the figure that one iteration consists of Mutation, Crossover and Selection process which is repeated for each vector. A mutant vector is created for each target vector in NP using Eq 4.1. In the next part, a trial vector is generated through the process of Crossover using Eq 4.2. As a last part of the iteration, the trial vector so created is compared with the target vector, i.e., their objective function values are compared. The vector, trial or target, having a better objective function value is retained. This is repeated for each and every vector in NP. Once all vectors have passed through Mutation, Crossover and Selection the new iteration will start, but before moving to the next iteration, the stopping criterion is checked. If the stopping criterion is not satisfied, a new iteration is started otherwise the results are displayed and the optimization process stops.

Figure 4.2 shows the flowchart for the creation of trial vector through the process of Crossover. While generating a trial vector, the loop repeats from 1 to 'D'. In each iteration, a random number is generated which is compared with CR and it is also verified if 'j' equates a predecided integer. In case this condition is satisfied, the mutant vector value of jth input variable is stored in the trial vector. For the input variables where condition is evaluated as false, the target vector value of the jth input variable is copied in the trial vector. This loop is used for every iteration in DE to generate the trial vector.

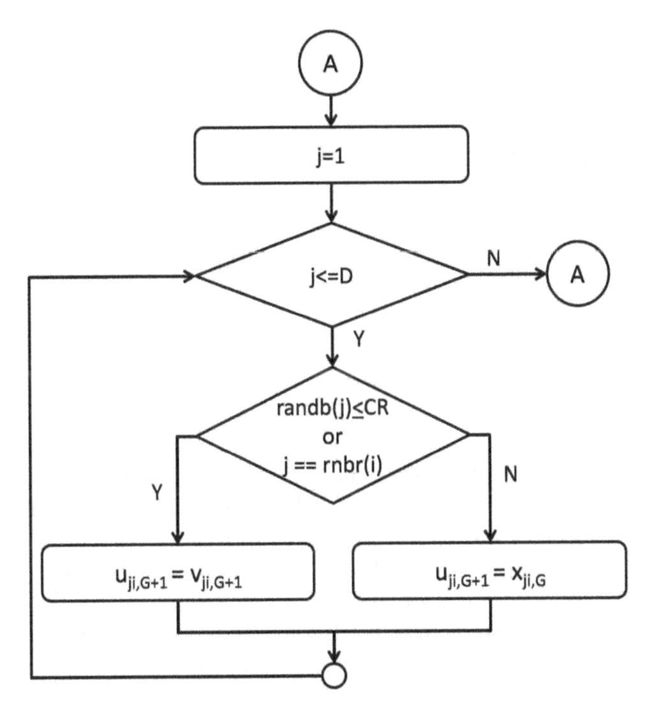

Figure 4.2: Trial Vector Generation Flowchart

4.6 EXAMPLE

DE is a simple optimization technique based on two equations and three processes. This section helps us understand the step-by-step execution of this optimization technique. Let us start with a set of randomly generated vectors. The total number of input variables are assumed to be 5 and total number of vectors, i.e., NP is taken as 6. Also the limits for input variables are to be kept at −600 to 600. In this example we would be applying the DE method for solving the Griewank problem which is defined by equations: 2.5 and 2.9. A sample solution is presented in 5.9.

A sample vector is represented as in Table 4.1.

The random NP vectors along with their values are presented in Table 4.2.

The objective function values for these vectors are found to be: 138.1064, 75.35377, 132.9249, 78.76276, 154.5424 and 120.6237, respectively.

Table 4.1
Sample Vector

	Variable 1	Variable 2	Variable 3	Variable 4	Variable 5
Sample Vector	16.0475	−277.517	74.94419	247.9742	391.221

Table 4.2
NP - Collection of Vectors

	Variable 1	Variable 2	Variable 3	Variable 4	Variable 5
Vector 1	141.8811	−104.656	−397.606	211.7688	561.0059
Vector 2	−16.0475	−277.517	74.94419	247.9742	391.221
Vector 3	503.4618	202.9188	51.23069	356.5671	321.7123
Vector 4	−80.9499	−329.007	−180.324	77.46379	397.8965
Vector 5	471.3352	−152.64	86.91336	386.1857	−460.428
Vector 6	−556.071	286.2077	118.7875	249.8144	106.6543

In this example, we would be taking a single fixed value for ‘CR’ and ‘F’, for simplicity and to make it understandable. The readers are encouraged to keep them dynamic i.e., generate them at the time of use, to observe the effects of it. We will have CR = 0.4402 and F = 1.2972. These values are randomly generated and it has been specially taken care of that these values are not extreme. Again it is up to the readers to observe the performance of DE for boundary values of these variables.

The iteration process should start with this much of data. We need to first generate a mutant vector for the 1st target vector i.e., i = 1. For generating a mutant vector, we need three random index numbers. The random numbers generated are: 5, 2 and 4 representing r_1, r_2 and r_3, respectively in Eq 4.1. Care has to be taken to ensure that these random numbers are not equal to each other and also that they should not be equal to ‘i’.

Thus Eq 4.1 can be written as:

$$v_{1,G+1} = x_{5,G} + F \cdot (x_{2,G} - x_{4,G})$$

After the generation of the mutant vector, it is checked for limit violation of input variables. In the above case, the 4th input variable crosses the limit and is thus restored to the maximum allowable limit of 600.

Once the mutant vector has been generated, the next step is to form the trial vector (u). For forming the trial vector, we would require: (a) rnbr(i) and (b) random numbers equal to the number of input variables. These random variables can be generated dynamically for each input variable. The five random numbers generated for five input variables are: 0.3588, 0.0633, 0.0494, 0.4841 and 0.45. Also the value of ‘rnbr(i)’ is generated as 3.

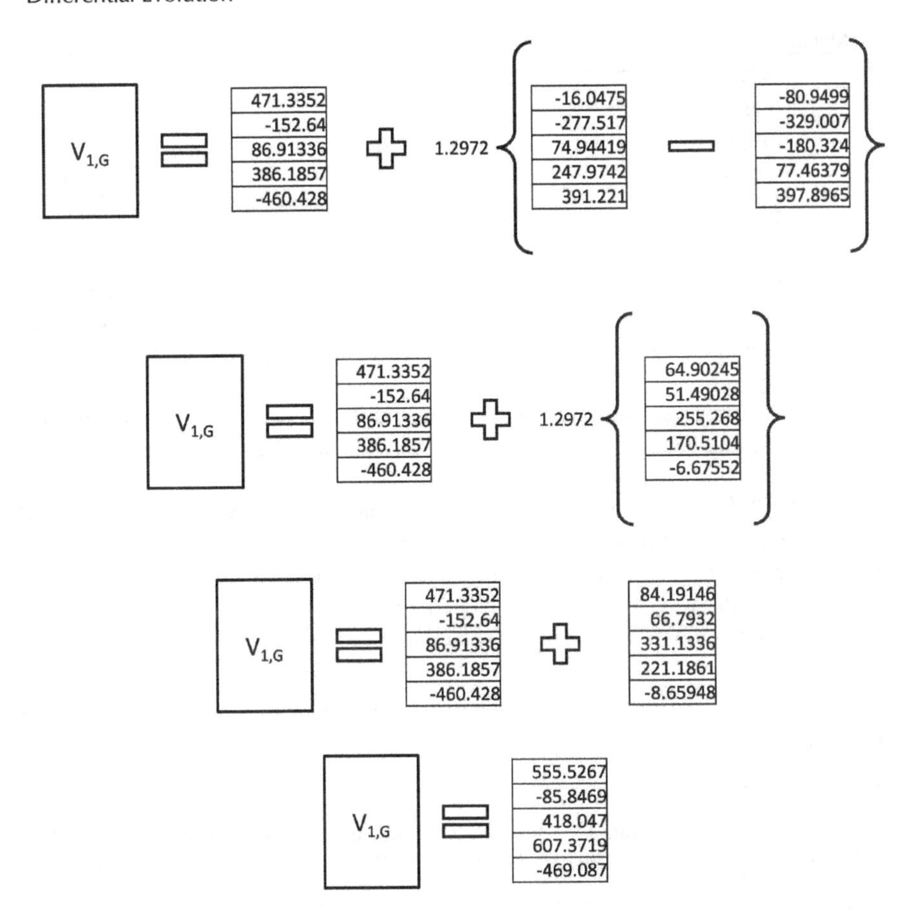

According to Eq 4.2, the trial vector shall be generated by visiting every input variable and deciding its value for the trial vector. If the respective random number of that input variable is less than CR or if the input variable index is equal to 'rnbr(i)' then the corresponding input variable value from the mutant vector is copied into the trial vector, otherwise the corresponding input variable value from the target vector is copied into the trial vector.

For first input variable the random number is 0.3588 which is less than 0.4402, i.e., CR. Thus the first input variable value from mutant vector will be copied in the trial vector. Similarly, we can observe that the random number corresponding to 2nd and 3rd positions are also lesser than CR. For the last two input variables, the random numbers are greater than CR and hence their values are copied from the target vector into the trial vector. Thus in the trial vector, the first three values are taken from the mutant vector and the remaining two values are taken from the target vector.

This is how the trial vector is generated. After forming the trial vector, its objective function value is evaluated, which in this case is found to be: 213.7931

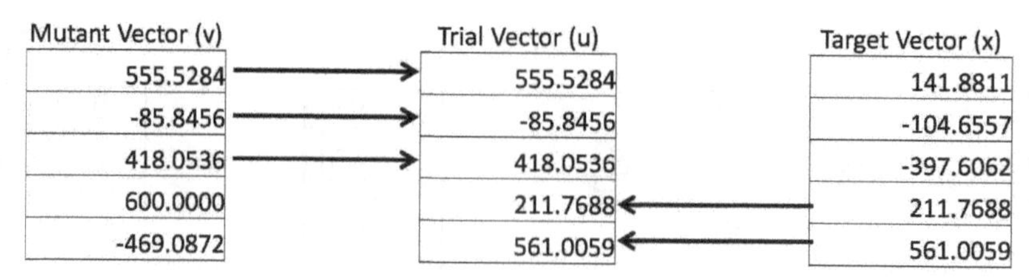

Since the objective function value of the trial vector (f(u)) is greater than objective function value of the target vector (f(x)), i.e., 213.7931 > 138.1064, the target vector is retained. This is the last process of selection for the first vector in the first iteration.

The three processes of mutation, crossover and selection shall be now repeated for the remaining vectors in NP in a similar way for the remaining part of the iteration.

For the 2nd vector, we have, $r_1 = 5$, $r_2 = 4$ and $r_3 = 3$. As we have observed earlier, a similar calculation using Eq 4.1 is performed for the new target vector. The reader is suggested to perform these calculations themselves to understand the process in a better way. The value of second input variable value goes out of limit and has to be fixed at the minimum limit. The mutant vector is presented in Table 4.3:

Table 4.3
Mutant Vector 2

	Variable 1	Variable 2	Variable 3	Variable 4	Variable 5
Mutant Vector2	−286.7789	−842.6681	−213.4651	24.1257	−361.5994

For crossover in the second target vector, the value of rnbr(i) is 4 and the random numbers generated are: 0.2848, 0.1807, 0.817, 0.1452 and 0.1426. Only 3rd random number is having a value greater than CR and thus the trial vector will be the same as the mutant vector, except that the 3rd value will be from the target vector.

Thus we have the second trial vector as given in Table 4.4:

Table 4.4
Trial Vector 2

	Variable 1	Variable 2	Variable 3	Variable 4	Variable 5
Trial Vector2	−286.7789	−600	74.9442	24.1257	−361.5994

The objective function value of trial vector 2 is found to be 145.8315. The target vector 2 has an objective function value of 75.3538. Since $f(u_2) > f(x_2)$, the target vector is retained.

For the third target vector, the calculations are summarized as:
Mutation:

r_1	5
r_2	4
r_3	6

Mutant Vector 3
1087.6743
-950.7126
-301.1016
162.6080
-82.6205

The trial vector generated for the third vector is given as:

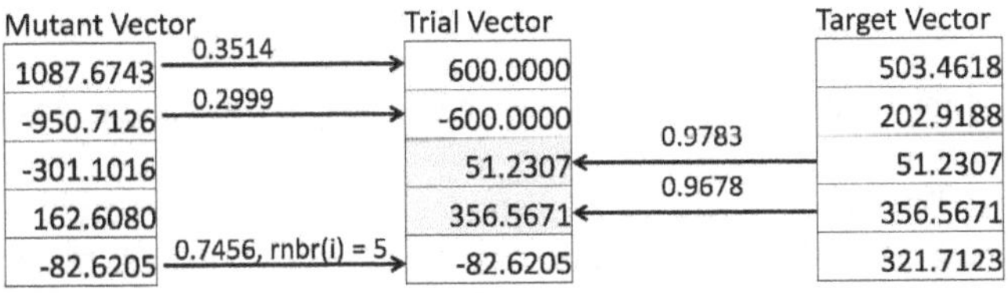

Mutant Vector	Random number	Trial Vector	Random number	Target Vector
1087.6743	0.3514 →	600.0000		503.4618
-950.7126	0.2999 →	-600.0000		202.9188
-301.1016		51.2307	← 0.9783	51.2307
162.6080		356.5671	← 0.9678	356.5671
-82.6205	0.7456, rnbr(i) = 5 →	-82.6205		321.7123

From the above calculation it can be clearly observed that the input variables with a random number less than CR are transferred from the mutant vector whereas the input variables with random numbers greater than CR are transferred from the target vector to form the trial vector. The only exception is that of the last input variable where the random number is greater than CR, but still the mutant vector value is transferred into the trial vector due to the rnbr(i) value being 5.

The objective function value of the new trial vector is found to 215.0131. The original target vector has an objective function value of 132.9249, and thus the target vector is retained.

It should be brought to the notice of the readers here that we have not been able to get a better trial vector until now, because of the NP size being small. With a bigger NP, may be the results would differ a little.

Let's move to the next target vector, the mutant vector is formed with $r_1 = 5$, $r_2 = 2$ and $r_3 = 3$.

The mutant vector is found to be:

-202.586
-775.874
117.6751
245.3162
-370.259

After correcting the limits we move to form the trial vector. The value of rnbr(i) comes out to be 4 and the random number values are: 0.8324, 0.585, 0.4326, 0.8149, 0.8262. Even though the 4th random value is greater than CR, because of rnbr(i) the 4th value in trial vector is copied from the mutant vector. Along with the 4th

input variable value, the 3rd input variable value is copied from the mutant vector. Remaining values are taken from the target vector.

The objective function value of the trial vector is found to be 87.6663. Again the trial vector objective function value is greater than the target vector objective function value, i.e., 78.7628. Thus the target vector is retained.

Hoping to find a better solution we move to the 5th target vector!!! The r_1, r_2 and r_3 values are taken as: 3, 1 and 2, respectively. The mutant vector comes out to be:

708.3309
427.1585
-561.774
309.6005
541.9617

The first input variable value has crossed the upper limit and needs to be fixed to 600.

The rnbr(i) value is generated as: 3 and the random numbers are: 0.7123, 0.6819, 0.4913, 0.5201, 0.6002. All the values are greater than CR and thus only the input variable corresponding to rnbr(i) is copied from the mutant vector to the trial vector. This serves as an example as to why the rnbr(i) term has been introduced. In its absence, the trial vector and target vector would be the same.

471.3352
-152.64
-561.774
386.1857
-460.428

The objective function value is found to be 231.5399, which is greater than the target vector objective function value of 154.5424. The target vector is therefore retained.

For the last vector in NP, we have r_1, r_2 and r_3 as 4, 1 and 3, respectively. The mutant vector is found to be:

-550.002
-728.001
-762.567
-110.372
708.3144

The 2nd, 3rd and 5th input variable values have crossed the specified limits and need to be adjusted to their respective limits.

For the crossover process, rnbr(i) = 5 and the random numbers are generated as: 0.3396, 0.2143, 0.0486, 0.5477, 0.9989. In this case all values in the trial vector are copied from the mutant vector except the 4th input variable value. The trial vector is:

-550.002
-600
-600
249.8144
600

The objective function value of the trial vector comes out to be 362.3565 whereas the objective function value of target vector is 120.6237. The target vector is therefore retained.

It can be observed from here that none of the trial vectors proved to be better than the corresponding existing target vectors. DE has these pitfalls in which the search process is not at all directed and the method heavily depends on random number generation. Also smaller size of NP, plays a spoil sport.

The values of all random numbers, mutant vectors and trial vectors for the next iteration are presented below. The reader is encouraged to perform the calculations and get a feel of the method.

It should be noted here that the calculations shall be done as shown in first iteration. Thus the reader should also move in a similar way and not perform the calculations in a parallel manner for all the vectors. In other words, one vector should go through the processes of Mutation, Crossover and Selection first and then move to the calculations for second vector and so on. On the other hand, do not perform Mutation for all vectors, simultaneously and then move to Crossover for all vectors and so on.

For the second iteration, the values of r_1, r_2 and r_3 are tabularized in Table 4.5.

Table 4.5
r_1, r_2 and r_3 for Iteration 2

	r_1	r_2	r_3
Vector 1	2	4	3
Vector 2	5	3	4
Vector 3	1	5	4
Vector 4	5	6	2
Vector 5	4	3	1
Vector 6	3	2	1

The mutant vectors so generated are tabularized in Table 4.6.

All the input variables crossing the limits have to be set to the maximum or minimum limits.

Moving on to the crossover state, the random numbers for rnbr(i) are: 5, 3, 3, 3, 4 and 4. The other set of random numbers generated are given in Table 4.7.

The trial vectors formed based on the above data are (Table 4.8):

The objective function values of these trial vectors are found to be: 111.4891, 204.2952, 197.1254, 137.831, 241.1489 and 174.7758. Compared to the

Table 4.6
Mutant Vectors for Iteration 2

	Variable 1	Variable 2	Variable 3	Variable 4	Variable 5
Mutant Vector 1	−774.162	−967.545	−225.434	−114.086	490.0491
Mutant Vector 2	1229.449	537.3878	387.2918	748.2458	−559.256
Mutant Vector 3	858.3198	124.1319	−50.9393	286.3963	−623.391
Mutant Vector 4	−229.197	578.6378	143.788	388.5728	−829.575
Mutant Vector 5	388.102	69.98672	401.9192	688.007	179.5256
Mutant Vector 6	298.5927	−21.3209	664.2354	826.2408	193.5098

Table 4.7
Random Numbers of Crossover for Iteration 2

	Variable 1	Variable 2	Variable 3	Variable 4	Variable 5
Vector 1	0.4535	0.8482	0.5987	0.1673	0.6183
Vector 2	0.4854	0.1034	0.9082	0.572	0.0366
Vector 3	0.7435	0.5512	0.1997	0.9103	0.225
Vector 4	0.0665	0.9778	0.9153	0.8472	0.2191
Vector 5	0.9508	0.0069	0.0723	0.5461	0.6481
Vector 6	0.8795	0.258	0.821	0.8199	0.5619

Table 4.8
Mutant Vectors for Iteration 2

	Variable 1	Variable 2	Variable 3	Variable 4	Variable 5
Trial Vector 1	141.8811	−104.656	−397.606	−114.086	490.0491
Trial Vector 2	−16.0475	537.3878	387.2918	247.9742	−559.256
Trial Vector 3	503.4618	202.9188	−50.9393	356.5671	−600
Trial Vector 4	−229.197	−329.007	143.788	77.46379	−600
Trial Vector 5	471.3352	69.98672	401.9192	600	−460.428
Trial Vector 6	−556.071	−21.3209	118.7875	600	106.6543

corresponding target vectors, only the first trial vector performs better than its target vector. Thus in NP the first vector is only replaced. Remaining vectors shall remain as it is.

Once the first trial vector is found to be better than its target vector, it should be replaced in NP. In further calculations, the new values of target vector 1, should be used.

The new NP is shown in Table 4.9 with vectors and their respective objective function values.

Table 4.9
NP after Iteration 2

	Variable 1	Variable 2	Variable 3	Variable 4	Variable 5	Obj Fn
Vector 1	141.8811	−104.656	−397.606	−114.086	490.0491	111.4891
Vector 2	−16.0475	−277.517	74.94419	247.9742	391.221	75.35377
Vector 3	503.4618	202.9188	51.23069	356.5671	321.7123	132.9249
Vector 4	−80.9499	−329.007	−180.324	77.46379	397.8965	78.76276
Vector 5	471.3352	−152.64	86.91336	386.1857	−460.428	154.5424
Vector 6	−556.071	286.2077	118.7875	249.8144	106.6543	120.6237

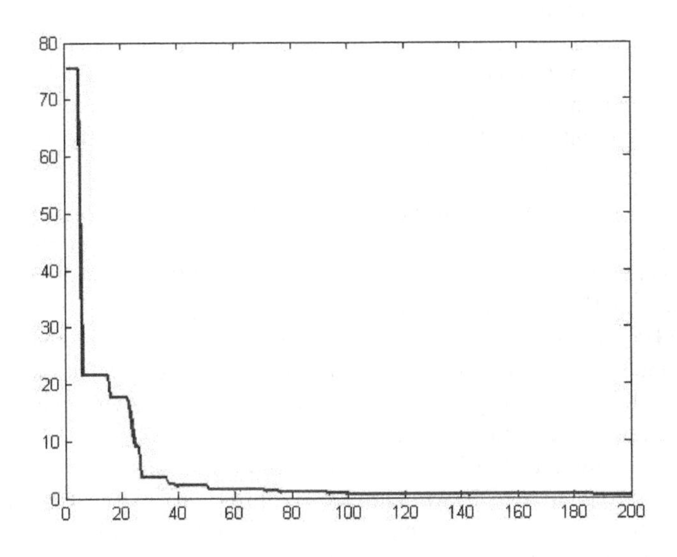

Figure 4.3: Variation of Best Vector over Iterations

Let us observe the change in the best vector objective function over the iterations for the same problem if executed for 200 iterations as shown in Fig 4.3. The final minimal value is 0.4596. However, the consistency of the output is not good in this method for these settings (like population size, etc).

4.7 VARIANTS AND HYBRID

This section is introduced to discuss the different variants and hybrid versions of DE. There may be multiple publications related to the same variant or hybrid, however, we would be discussing a single publication for the convenience of the reader and to get the concept clarified.

4.7.1 VARIANTS

Variants are defined as those versions of the process which proposes changes in the basic working of the optimization method.

1. **Recombination/Mutation Operator**: A variation in the combination and mutation operators is studied in [93]. The research paper presents eight variants with different methods of obtaining the combination and mutation operators. The variations are applied on 13 benchmark problems. A list of these variations is given below:

    rand/p/bin
    rand/p/exp
    best/p/bin
    best/p/exp
    current-to-rand/p
    current-to-best/p
    current-to-rand/p/bin
    rand/2/dir

    These variations are considered as strategies in other research work. It is left to the readers to go into the details of implementing these variations. A similar collection of variations in DE is presented in [90, 172]. Another strategy of "DE/rand/1/bin" is presented in [17], termed as jDE.
2. **JADE**: A mutation strategy is implemented to control the optimization method termed as JADE [185]. The constants of F and CR are controlled adaptively and "DE/current-to-pbest" with optional archive is applied. The method is applied on 13 benchmark problems.
3. **Composite DE (CoDE)/Strategy Adaptation based JADE (SaJADE)**: Composite DE (CoDE) was introduced in [169]. In CoDE, different trial vector generation strategies are randomly combined through various control parameter settings, in the formation of trial vectors for each generation.
    In SaJADE [49], a similar methodology is applied. It is considered also as an improved version of JADE, thus the name SaJADE. The strategies are selected through the Eq 4.3

    $$S_i = [\eta_i \times K] + 1 \tag{4.3}$$

    where S_i is the mutation strategy to be implemented

$\eta_i \in [0,1]$ is strategy parameter, used for selection of strategy.
K is the set of all mutation strategies.

4. **Self-adaptive DE (SaDE)**: Self Adaptive DE (SaDE) collects a set of strategies to generate trial vectors [123]. These strategies are kept as a set to be chosen at run time and have diverse characteristics. During the generation of trial vector, one strategy is chosen from the set of strategies. The strategy chosen is based on the probability associated with it. This probability is dependent on the performance of the strategy in the previous generation.
5. **Self-adaptive neighborhood Search DE (SaNSDE)**: Self-adaptive neighborhood Search DE (SaNSDE) [178] is a combination of SaDE and neighborhood Search DE (NSDE). The combination occurs on three fronts: mutation strategies for self adaptation, self adaptation of F and CR. For the first and third part related to mutation strategies and CR self adaptation, SaDE is followed while in the self adaptation of F, NSDE is followed.
6. **Self Adaptive Pareto DE (SPDE)**: The Self Adaptive Pareto DE (SPDE) in [2] introduces a DE variant in which the operators are self adaptive based on pareto DE. Some of the rules in this variant are: variables shall lie between 0 and 1, initial population is generated randomly with a mean of 0.5 and standard deviation of 0.15 for a Gaussian distribution, dominated solutions shall not be a part of the population and three parents shall be utilized to form a child.
7. **Fuzzy Adaptive DE (FADE)**: As with the above presented variants, the basic idea of controlling and obtaining optimum F and CR is presented in [84] through the use of Fuzzy Logic.
8. **Opposition Based DE**: The idea of opposition based learning is mixed with DE in [125]. The vectors are generated randomly and along with them the opposition based vectors are also generated. The vectors to be retained are then selected through the greedy criterion. The opposite of each variable is evaluated and thus the opposite vector is dynamically created.
9. **Quantum DE**: N Queen problem is taken as reference in [34]. Quantum DE is proposed to save the time expended in solving the problem. In Quantum DE, one vector encodes more than one solution. The process has two parts with the first working on quantum bit and unitary quantum operator while in the second part, quantum differential mutation is used. The vectors in this variant are termed as quantum vectors/chromosomes.
10. **Cooperative DE**: A method to solve multiobjective optimization problems is presented in [167]. The method has multiple populations to solve the multiobjective problem which is defined as a collection 'M' single objective functions. The number of populations is one more than the number of single objective functions. The 'M' populations try to optimize the 'M' objective functions while the archive population maintains the non dominated solutions and guides the other populations to search along the pareto front. The method uses modified JADE for mutation of 'M' populations while the archive population utilizes jDE.
11. **Binary DE**: Trigonmetric function and angle modulation is used in [107] to apply DE for a binary space. The equation used for angle modulation is given

by 4.4.

$$g(x) = \sin(2\pi(x-a) \times b \times \cos(A)) + d \tag{4.4}$$

where

$$A = 2\pi \times c(x-a)$$

a, b, c and d are constants
and x is one of the intervals in the binary space.

12. **Memetic DE**: In [155], a combination of DE, Hooke Jeeves Algorithm and Stochastic Local Searcher is presented as memetic DE. The local search method helps DE to overcome the problem of stagnation. Initially, the vectors are evolved according to the original DE and then after a certain number of evaluations, local searches are activated based on the value of parameter 'v'. This parameter is evaluated as in Eq 4.5.

$$v = \min\left\{1, \frac{\sigma_f}{|f_{avg}|}\right\} \tag{4.5}$$

13. **Bare Bones DE**:The variant of Bare Bones DE was introduced in [104]. The equations utilized in the variant are Eqs 4.6 and 4.7.

$$x_{ij}(t) = \begin{cases} p_{ij}(t) + r_{2j} \times (x_{i_1,j}(t) - x_{i_2,j}(t)) & \text{if } U(0,1) > p_r \\ y_{i_3,j}(t) & \text{otherwise} \end{cases} \tag{4.6}$$

$$p_{i,j}(t) = r_{1,j}(t) y_{i,j}(t) + (1 - r_{1,j}(t))\, \hat{y}_{i,j}(t) \tag{4.7}$$

where $x_{ij}(t)$ is the new position of the vector under consideration;
i_1, i_2, i_3 are index numbers within the population and $i \neq i_1 \neq i_2$.
p_r is recombination probability
r_{1j}, r_{2j} are random numbers between 0 and 1.
Bare Bones DE introduces p_i for attraction of vector towards personal and global best, such that it acts as a stochastic weighted average of both. Crossover is also implemented through y_{i_3} which is a randomly selected personal best of any of the vectors within the population size.

14. **DE for Multi-Objective (DEMO)**: The application of DE for optimizing multiobjective problem is discussed in [133]. The paper explains the method through algorithms. The vector generated for the next iteration is retained if it dominates the vector from which it was generated.
15. **Unified DE (UDE)**: UDE [158] combines [123], [169] and [185] along with some other DE Variants. UDE works on a collection of three trial vector generation strategies and two control parameter settings. After each iteration, the population is divided into two parts. For the better part of the population all the trial vector generation strategies are applied. For the other part of the population, the trial vector strategies are self adapting, based on their performance in generating superior solutions in the better part of the population. The parameter pools are: [F=0.9, CR=0.9] and [F=0.5, CR=0.5].

4.7.2 HYBRID DE

A combination of DE and some other optimization techniques is considered as a hybrid optimization technique and such techniques are covered in this section. As can be observed from the previous section, even after covering a large number of variants of DE, some variants could not be accomodated. A similar scene but infact a more exaggerated scene can be viewed for the compilation of hybrid DE techniques. Thus a brief compilation of DE hybrids is presented in Table 4.10. The readers are suggested to understand the basic concept of proposed hybrid techniques in the literature and not the problem on which it is applied.

Table 4.10
Hybrid PSO

Sr. No.	Hybrid With	Reference
1	Genetic Algorithm	[54]
2	Particle Swarm Optimization	[28, 70, 103, 109]
3	Ant Colony Optimization	[124]
4	Chaotic	[20, 23, 33, 73, 86, 87, 114]
5	Artificial Bee Colony	[60, 76, 78, 157, 174, 190]
6	Harmony Search Algorithm	[131]
7	Simulated Annealing	[119]
8	Tabu Search	[85]
9	Gravitational Search Algorithm	[77]
10	Sequential Quadratic Programming	[36, 54]
11	Immune Algorithm	[81]
12	Taguchi	[180]

4.12 HYBRID DE

A combination of DE and some other optimization techniques is [illegible] optimization techniques and such techniques are covered [illegible] has been [illegible] such as even altered by a large number [illegible] [illegible] the hybridization of DE [illegible] [illegible] population of hybrid DE [illegible] Table 4.10. The [illegible] [illegible] of proposed hybrid techniques [illegible] and the application [illegible]

Table 4.10
Hybrid DE

[illegible]	[illegible]	References
[illegible]	[illegible]	[illegible]
1	Particle Swarm Optimization	[28, 29, 101, 102]
[illegible]	Ant colony optimization	[illegible]
[illegible]	[illegible]	[20, 23, 33, [illegible]]
[illegible]	Artificial Bee Colony	[100, [illegible]]
[illegible]	[illegible]	[illegible]
[illegible]	[illegible]	[illegible]
[illegible]	[illegible]	[illegible]
[illegible]	[illegible]	[illegible]
[illegible]	[illegible]	[illegible]
[illegible]	[illegible]	[illegible]
[illegible]	[illegible]	[illegible]

5 Particle Swarm Optimization

5.1 INTRODUCTION

Particle Swarm Optimization (PSO) is one of the most widely applied EOAs. It is robust and can easily be implemented on a wide variety of problems. PSO imitates the behavior of flocks of birds or pools of fish. The movement of the flock of birds or school of fish is directed towards optimization of food search, security from predators or for other survival reasons. The movement of these animals, as shown in Fig 5.1, is optimized based on their coordinated movements. Just like the birds are able to fly much longer distances when they fly in a certain alignment, the fish can divulge attacks from predators through their coordinated movements.

There is, however, one drawback in the movements of birds and fish and that is, two birds or fish cannot occupy the same physical space, simultaneously. It can be reciprocated in PSO by not allowing two particles to occupy the same space position in search space. But this is not a precondition in PSO. Let us, therefore, compare PSO with a human's thinking and decision making process. Human decision making capabilities are dependent on factors like social and cognitive behavior. These behaviors take into consideration an individual's own experience and the experience of the society. It is also possible that the thoughts of two persons are the same as shown in Fig 5.2, where two individuals can be thinking of the same car at the same time. Thus the thoughts which do not occupy physical space can be replicated over different individuals, which is a possibility in PSO.

PSO was introduced by Kennedy and Eberhart in 1995 in [67]. PSO has since its inception seen a variety of changes starting with James Kennedy [68] revisiting PSO in 1997, exploring the possibility of implementing human behavior in PSO for reaching an optimized solution for a given problem. The concept of individual best and neighborhood best (i.e., the global best) is also discussed. Various human behavior traits like cognition only, social only and selfless models are corelated with PSO.

One of the advantages of PSO, as far as coding is concerned is that it does not require to continuously compare recently calculated positions with the previously obtained positions, instead it is just required to track of the global best. This also reduces the requirement of variables in the program. In fact, PSO operates on two simple equations only.

For a very brief description, PSO is an iterative EOA, having a swarm of particles searching the search space. The particles move in the search space based on their velocities. The velocity of a particle is dependent on three components: inertia, its own best position during the iterative process and best position attained by any of the particles in the swarm.

DOI: 10.1201/b22647-5

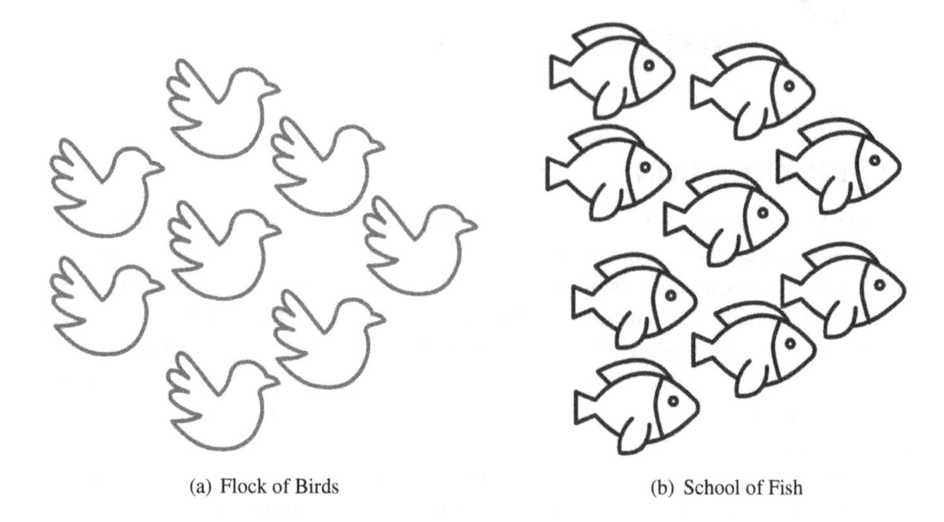

Figure 5.1: Concept of Particle Swarm Optimization

Figure 5.2: The Same Thoughts of Individuals

One of the downsides to PSO is that the particles are never regenerated. Thus, if the initial population which is generated randomly is concentrated in a local area within the search space, then the optimization process shall converge to the local optima of the local region in which the particles are present.

The changes and hybrids introduced in PSO are discussed in Section 5.8.

5.2 TERMINOLOGY

This section presents the definition and mathematical representation of different terms used in PSO, with a little help from [4]:

1. **Particle X(t):** A particle represents a possible solution to the optimization problem having an n-dimensional vector, where n is the number of parameters or input variables which are to be optimized. The ith particle, given by X_i can be presented as $X_i = [x_{i,1}, x_{i,2}....x_{i,n}]$, where i gives the particle number in the swarm, and 1,2,....n are the input variables. It would further be very clear once the reader goes through the Section 5.7, with a sample particle shown in Table 5.1.
2. **Population:** A collection of *'p'* particles is a population. The population can be represented as P(t) = $[X_1(t), ...X_p(t)]$.
3. **Swarm:** A Swarm is a population in which the particles are moving randomly to finally cluster together, i.e., converge, at the end. The Swarm is represented in Table 5.2.
4. **Particle Velocity V(t):** It is the velocity of moving particles represented by an n-dimensional vector. At time t, the jth particle velocity $V_j(t)$ can be described as $V_j(t) = [v_{j,1}(t), ...v_{j,n}(t)]$, where $v_{j,k}(t)$ is the velocity component of the jth particle with respect to the kth dimension. The size of velocity vector of a particle will be the same as that of the particle itself.
5. **Inertia Weight ω(t):** It is used in calculation of velocity and controls the effect of previous velocity on current velocity. It is responsible for the optimization technique searching globally or locally. Usually in the initial stages of the search process, large inertia weight is used for global exploration while, the inertia weight is reduced in later stages for local exploitation. It is proposed to change in relation with the iteration count.
6. **Individual Best $X^*(t)$:** The particle moves during the execution of PSO within the search space. In this movement of the particle, the objective function value is evaluated for each new position that the particle visits. The position where the particle attains the best objective function value is stored as its Individual Best. The individual best $X^*(t)$ is determined and updated for each particle in the swarm for every iteration.
7. **Global Best $X^{**}(t)$:** The best position out of all the individual best positions is the global best. Global best position has the optimal objective function value achieved by any particle until a given iteration. Global best shall represent the optimal solution, in case the particles do not converge at the end of the search process.

8. **Stopping Criteria:** Stopping criteria are a set of conditions under which the search process of PSO can be terminated. The search process in PSO can terminate if one of the following criteria is satisfied: (a) the number of iterations since the last change of the best solution is greater than the pre-specified number of iterations or (b) the number of iterations reaches the maximum allowable number or (c) all the particles have converged to a single optimal position.

5.3 EVOLUTION OF PARTICLE SWARM OPTIMIZATION

PSO was introduced in 1995 by Kennedy and Eberhart [67]. Regular changes were proposed over the years in PSO, which is discussed in this section; however, only the contributions by Kennedy, Eberhart and Shi are presented here, for better understanding and to follow a linked progress. PSO has therefore evolved in a manner as shown through the equations below.

A particle in PSO can be given as $X_i = (x_{i1}, x_{i2}, ..., x_{in})$ for a n-dimensional search space. As stated earlier, the position of a particle changes every iteration based on individual best position and global best position.

The calculation of velocity plays a very important role in convergence of particles in PSO. Also all the variations that are introduced are in the evaluation of velocity. Initially, an inequality test was employed in [67] to find velocity adjustments as shown in Algorithm 5.1:

Algorithm 5.1 Velocity Adjustments in PSO

if $presentx > bestx$ **then**
  $V_x = V_x + random * g\_increment$
else if $presentx < bestx$ **then**
  $V_x = V_x - random * g\_increment$
end if

where g_increment is a system parameter.

This inequality test was actually very rigid and is unable to vary the speed of the particles. The movement of the particles will be very restricted. Also the probability of particles getting stuck in local optima is quite high. Thus the inequality test was done away with. In the same research work, therefore, velocities for each dimension was found through acceleration based on the distance of the particles from the best position as given in the equation below:

$$vx[\ ][\ ] = vx[\ ][\ ] + rand() * p\dot{}increment(pbestx[\ ][\ ] - presentx[\ ][\ ]) \quad (5.1)$$

where $vx[\ ][\ ]$: Velocity of a particle
$p\dot{}increment$: Function defining the difference between global or individual best and current position of the particle.
pbestx : Global or Individual best.
presentx : Current position of the particle.
rand(): Random Value.

$vx[\ ][\ ]$ on the right hand side of the equation indicates, the velocity of the particle in the previous iteration.

The last phase of [67] ends with an equation, which makes p- or g- increment larger. The stochastic factor is multiplied by 2 so as to achieve a mean of 1, so that particles would "overfly" the target about half the time. In 1995, Kennedy and Eberhart were not sure whether the optimum value of the constants should be 2, or their value should be evolved for each problem or from some knowledge base of a particular problem. The formula for obtaining the velocities was then fabricated as:

$$\begin{aligned} vx[\ ][\ ] = vx[\ ][\ ] &+ 2*rand()*(pbestx[\ ][\ ] - presentx[\ ][\ ]) \\ &+ 2*rand()*(pbestx[\ ][gbest] - presentx[\ ][\ ]) \end{aligned} \tag{5.2}$$

where pbestx[ ][ ] : Individual best.
pbestx[ ][gbest] : Global best.

In 1997, James Kennedy [68], presented the equation 5.2 in terms of human behavior. The second term in equation 5.2 was considered as a representative of cognition model of human behavior whereas the third term represents the social model. The cognition model deals with the ability of a human to remember his past experiences whereas the social model considers the experience of others in the society.

The equation used for cognition only model is:

$$V_{id} = V_{id} + \phi_{id}(P_{id} - X_{id}) \tag{5.3}$$

where V_{id} : Velocity of a particle
X_{id} : Current position of a particle
P_{id} : Individual best.

When neighborhood particles are also considered and social model is introduced in the equation, then it can be represented as [68]:

$$V_{id} = V_{id} + \phi_{id_1}(P_{id} - X_{id}) + \phi_{id_2}(P_{gd} - X_{id}) \tag{5.4}$$

where P_{gd} : Global Best

The difference between the individual best and the particle position as well as global best and the particle position, is weighted by a positive random number ϕ, whose upper limit is considered as a parameter of the system. It was also suggested that the value of V_{id} be limited a particular value V_{max}. This limit would serve three purposes: (a) it keeps the computer from overflowing, (b) it realistically simulates the incremental changes of human learning and attitude change, and (c) it determines the granularity of search of the problem space.

The limit can be applied as [68]:

$$V_{id} > V_{max}, \text{ then } V_{id} = V_{max},$$
$$\text{and}$$
$$V_{id} < -V_{max} \text{ then } V_{id} = -V_{max}$$

The individual best and global best terms were now properly in place representing individual and social aspects of a human being. The next aspect of research was the ability of PSO to control the velocity of the particles. By controlling the movement of particles to search globally or locally, which is very important in a search process. In this context, globally means exploration and locally means exploitation. In exploration, a large area in search space is covered and the movement of particles will have bigger steps whereas in exploitation, the steps are small and the particles start searching the global optima in a localized area.

In 1998, Shi and Eberhart [142] introduced constants c_1 and c_2 in the velocity equation in place of ϕ_{id_1} and ϕ_{id_2}, respectively. The equation was thus modified as:

$$v_{id} = v_{id} + c_1 * rand() * (p_{id} - x_{id}) + c_2 * Rand() * (p_{gd} - x_{id}) \tag{5.5}$$

where c_1 and c_2 : Two positive constants.
rand() and Rand() : Two random functions within the range of 0 & 1.

Another term called inertia weight ω was then introduced in [142], [143]. The purpose of inertia weight is to balance the global and local search. Equation 5.5 was modified as:

$$v_{id} = \omega * v_{id} + c_1 * rand() * (p_{id} - x_{id}) + c_2 * Rand() * (p_{gd} - x_{id}) \tag{5.6}$$

where ω is the inertia weight.

ω and 'w' have been interchangeably used in different research papers.

If the inertia weight is large it will allow a global search, i.e., exploration while a small inertia weight will allow a local search, i.e., exploitation. In [143], optimal range of ω is found out to be 0.8–1 whereas in [142], the range of ω is found out to be 0.9–1.2. However, in both the research papers, it is suggested that the value of ω should be varying throughout the search process. In [142], it is further added that the value of ω should be time decreasing. A balance between global and local search is established by a proper selection of value for ω. It can have a positive constant value or a positive linearly varying value or a nonlinear function of time or a time decreasing value. A large value of ω ensures global search whereas a smaller value of ω helps to localise the search process. Thus for a global search initially and a local search later in the PSO search process, the inertia weight should linearly decrease from a large value to a small value through the course of PSO execution. Shi and Eberhart [141] suggests that the inertia weight should start with a value close to 1 and linearly decrease to 0.4 through the search process.

Another change in the calculation of velocity is brought in though the Constriction factor (K) in [22]. The constriction factor is defined in [35] as:

$$v_{id} = K * [v_{id} + c_1 * rand() * (p_{id} - x_{id}) + c_2 * Rand() * (p_{gd} - x_{id})] \tag{5.7a}$$

$$K = \frac{2}{|2 - \phi - \sqrt{\phi^2 - 4\phi}|} \tag{5.7b}$$

where $\phi = c_1 + c_2, \phi > 4$
K : Constriction factor.

The values of the constants c_1 and c_2 are analyzed in [35]. It was concluded that the optimum value of these constants is 2.05. Also the optimum value of K is therfore 0.729.

The equations (5.7a and 5.7b) have since then gone through a number of changes but this method of velocity calculation has been taken as a base for any further development. Equation 5.7a is taken as the final equation developed in the evolution of PSO.

5.4 FUNDAMENTAL CONCEPT

PSO is a robust optimization technique introduced by Shi and Eberhart, much of which has been discussed in earlier sections. The PSO method basically works on the concept of flocks of birds or school of fish. As explained in the Introduction section, a particle can also imitate the thought process of a human being. This section shall be kept small as the discussion on velocity is already taken care of in the previous section.

In PSO, the particles are initially generated within the search space randomly. A single particle is a candidate to an optimal solution. Each particle shall contain one representative data for each input variable. An example of a swarm and particles is given in Fig 5.3. The particles are spread in the search space. Each particle has

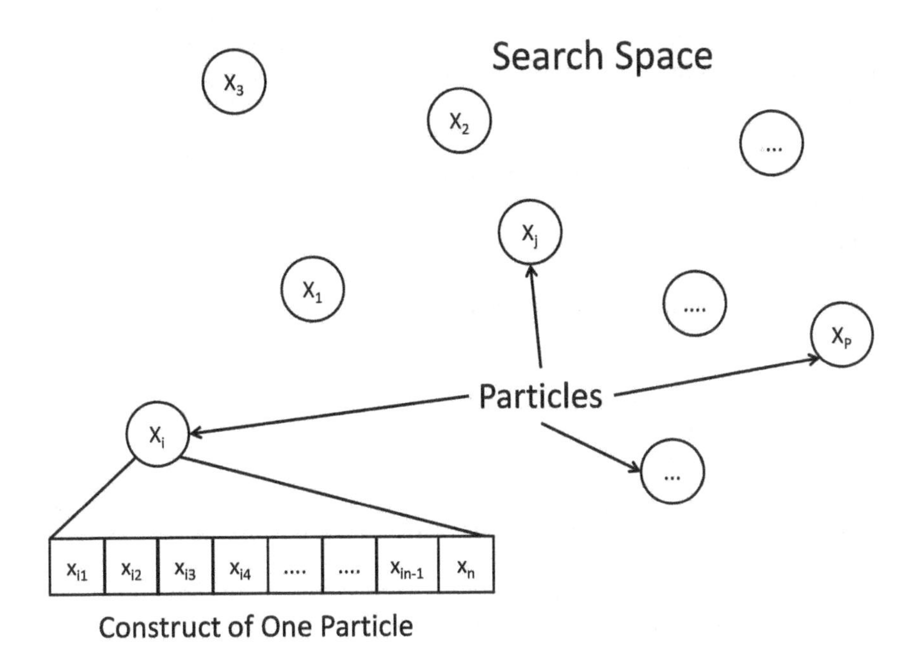

Figure 5.3: Particles and Swarm

one value representing an input variable as shown in the figure by x_{i1}, x_{i2},.....x_{in}. The particles are supposed to move in the search space and explore it, to find the optimal solution. To move in the search space, the particles are dependent on the velocity. The velocity componenet is already discussed in the previous section.

For each iteration the particle shall move in the search space and the new position of a particle is given as Eq 5.8:

$$x_{id} = x_{id} + v_{id} \tag{5.8}$$

The velocity variable shall have the same dimensions as the particle. The respective values of various dimensions in velocity and the particle position are added. It should be noted that the velocity components can be negative too. This addition is similar to vector addition. Lastly, the limits for individual dimension should be checked after the execution of the above equation and if some values have violated the limits then their values should be properly adjusted.

The operation of PSO is based on the execution of two equations in each iteration for each particle. These equations are velocity calculation and an update of particle position. This process shall continue until the stopping criterion has been met.

5.5 ALGORITHM AND PSEUDOCODE

PSO is one of the easiest optimization techniques as has been evident until now. In this section we shall view the algorithm for PSO. Slightly different implementations of PSO are discussed, in this section and in the next section, which will be clear to the readers as they read both the sections. Let's go through the algorithm of PSO in Algorithm 5.2.

Algorithm 5.2 Particle Swarm Optimization

```
START
for 1 to Swarm Size do
    Generate Particle
    if Variable > Max Limit or Variable < Min Limit then
        Variable = Limit
    end if
    Evaluate Objective Function Value
    Update Individual Best
    Update Global Best
end for
for 1 to Maximum Number of Iterations do
    for 1 to Swarm Size do
        Evaluate Velocity of the Particle
        Evaluate Objective Function Value
        Update Individual Best
        Update Global Best
    end for
end for
Display Result
STOP
```

As discussed earlier, PSO works only on two equations: the first one is required for the calculation of velocity and the second one is for the updating of particle position. In the algorithm, we generate random particles within the seach space, initially. The objective function value of the particle is simultaneously evaluated. Since the particles have not moved yet, the current position of a particle will be its individual best position. Also the global best is found in the same process.

Next we start the search process and loop from 1 to the maximum number of iterations. In each iteration, nest another loop which iterates up to the swarm size. Inside the inner loop, we find the velocity of the respective particle, then update its position and evaluate its objective function value. In the next part, the individual best of the particle is updated, in case a better solution is found. Also simultaneously the global best of the swarm is updated. The loops shall be repeated until the stopping criterion for PSO is met.

5.6 FLOWCHART

The flowchart for PSO is given in Fig 5.4. The flowchart of PSO is one of the easiest in all the optimization techniques. The loops for individual particles are not shown in the flowchart.

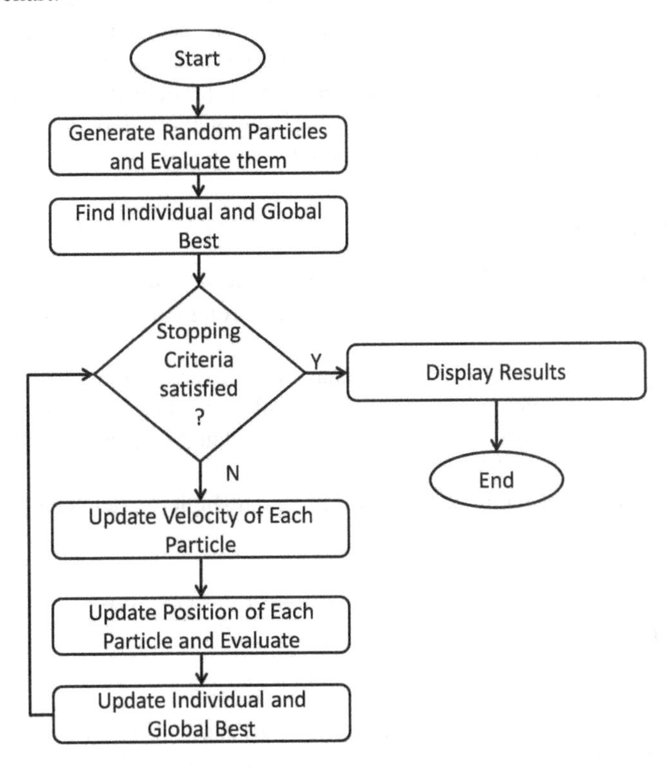

Figure 5.4: Flowchart - PSO

In PSO, the particles are randomly generated within the search space, initially. These particles are similar to the chromosomes in GA, except that each value in a particle represents a different dimension. After generation of particles, i.e., the swarm, the objective function value for each particle is evaluated. For the initially generated particles, the initial particle position itself is its individual best. The particle with the best objective function value gives the global best for the swarm. Next we enter the iteration mode of PSO, in which we have four specific steps. In the first step, the velocities for each particle is obtained using the Eq 5.7a. In the next step, we update the position of the particles using Eq 5.8. Once the particle positions are updated, the objective function value for each particle is found and the individual best positions of each particle and global best position of the swarm is updated. Next we check the stopping criterion, if it is satisfied, the iteration process will stop, otherwise the iterations will keep repeating.

A simple note should be made here: if the above flowchart is followed, then we can have individual loops for each task inside the bigger outside loop which shall have the stopping criterion condition. In other words, first the velocities of 'all' the particles shall be calculated, then their positions will be updated, then their individual best positions shall be updated and in the end the global best position will be found. Else a small variation can be introduced as shown in Alg 5.2, in which inside the outer loop for stopping criterion, we can have a loop encompassing all particles, i.e., from 1 to swarm size, in which we perform all tasks for one particle 'at a time'. In other words, the inner loop will find velocity for one particle, then update its position, then update its individual best and if needed the global best too.

This loop shall be repeated for all particles.

5.7 EXAMPLE

In this section, we will solve the PSO algorithm step-by-step and see how the values of different variables are calculated and updated throughout the search process during different iterations.

Let us assume that the number of input variables are 5 in number, i.e., n = 5 and the optimization problem is Griewank. For details related to the solving the Griewank problem refer equations 2.5 and 2.9. The PSO related calculations will performed using equations 5.7a and 5.8.

If the number of input variables are 5, then a single particle can be given as in Table 5.1:

Table 5.1
Sample Particle

	Variable 1	Variable 2	Variable 3	Variable 4	Variable 5
Sample Particle	16.0475	−277.517	74.94419	247.9742	391.221

Let us assume an initial sample swarm population to be as given in Table 5.2: It

Table 5.2
Sample Swarm

	Variable 1	Variable 2	Variable 3	Variable 4	Variable 5
Particle 1	141.8811	−104.656	−397.606	211.7688	561.0059
Particle 2	−16.0475	−277.517	74.94419	247.9742	391.221
Particle 3	503.4618	202.9188	51.23069	356.5671	321.7123
Particle 4	−80.9499	−329.007	−180.324	77.46379	397.8965
Particle 5	471.3352	−152.64	86.91336	386.1857	−460.428
Particle 6	−556.071	286.2077	118.7875	249.8144	106.6543

should be kept in mind while generating the particles that the variable values should be within the specified range. In this problem, the allowable range of each variable is from −600 to 600. The current size of the sample swarm is taken as 6 for better display and understanding of the calculations. The particles are generated randomly within the search space. It can be seen from Table 5.2 that each particle contains 5 variables each. The next step would be to find the Objective Function value of each particle.

Let us solve the Griewank Equation for one solution, like third particle:

$$f(x) = 1 + \left(\frac{x_1^2}{4000} + \frac{x_2^2}{4000} + \frac{x_3^2}{4000} + \frac{x_4^2}{4000} + \frac{x_5^2}{4000}\right)$$
$$- \left(cos\left(\frac{x_1}{\sqrt{1}}\right) * cos\left(\frac{x_2}{\sqrt{2}}\right) * cos\left(\frac{x_3}{\sqrt{3}}\right) * cos\left(\frac{x_4}{\sqrt{4}}\right) * cos\left(\frac{x_5}{\sqrt{5}}\right)\right)$$
$$= 1 + \left(\frac{503.46^2}{4000} + \frac{202.92^2}{4000} + \frac{51.23^2}{4000} + \frac{356.57^2}{4000} + \frac{321.71^2}{4000}\right)$$
$$- \left(cos\left(\frac{503.46}{\sqrt{1}}\right) * cos\left(\frac{202.92}{\sqrt{2}}\right) * cos\left(\frac{51.23}{\sqrt{3}}\right) * cos\left(\frac{356.57}{\sqrt{4}}\right) * cos\left(\frac{321.71}{\sqrt{5}}\right)\right)$$
$$= 1 + (63.37 + 10.29 + 0.66 + 31.79 + 25.87) - (0.69 * 0.52 * -0.26 * -0.71 * 0.8)$$
$$= 1 + 131.9783 - 0.0534 = 132.9249 \quad (5.9)$$

As the number of input variables are 5, the equation is also readjusted for 5 terms in summation and product parts of the equation.

The objective function is thus evaluated for all the particles in the swarm. For the swarm taken in Table 5.2, the objective functions are evaluated as given in Table 5.3.

For the first iteration, the Individual Best shall be the same as that of the Swarm. The individual best shall be found from the next iteration onwards. Thus the individual best shall be given by Table 5.2. It must also be noted that the dimensions of the individual best matrix should be the same as that of the swarm, since each particle will have its own individual best.

Table 5.3
Objective Functions for the Sample Swarm

	Objective Function Value
Particle 1	138.1064
Particle 2	75.35377
Particle 3	132.9249
Particle 4	78.76276
Particle 5	154.5424
Particle 6	120.6237

Global best will be the particle which has the minimum objective function value within the Swarm. From Table 5.3, it can be observed that Particle 2 has the lowest value, i.e., 75.35377, for its objective function. Thus the global best will be the same as that of Particle 2. The dimension of Global best matrix should be the same as that of a single particle.

To find the velocity for each particle, let us first generate a random matrix. It is not compulsory to generate such a matrix, as random values can be dynamically generated while calculating velocities for each variable. Dynamically generated random numbers will also help us have different random numbers for individual component and global component of velocity calculation. Generation of random matrix shall help us evaluate the calculation of the velocity matrix. The dimension of random matrix should be the same as that of a Swarm. The random matrix should have values between 0 and 1. For the first iteration the random matrix being considered is given in Table 5.4:

Table 5.4
Random Numbers for Velocity Calculation - Iteration 1

	Variable 1	Variable 2	Variable 3	Variable 4	Variable 5
Particle 1	0.07594	0.5688	0.3112	0.6892	0.1524
Particle 2	0.054	0.4694	0.5285	0.7482	0.8258
Particle 3	0.5308	0.0119	0.1656	0.4505	0.5383
Particle 4	0.7792	0.3371	0.602	0.0838	0.9961
Particle 5	0.934	0.1622	0.263	0.229	0.0782
Particle 6	0.1299	0.7943	0.6541	0.9133	0.4427

Let us now see, how to calculate the velocity of a particle. The velocity of the particle is given by Eq 5.7a. The following Eq 5.10, shows the calculation of velocity for third variable of Particle 5.

$$v_{53} = 0.729 * [0 + 2.05 * 0.263 * (86.91 - 86.91) + 2.05 * 0.263 * (74.94 - 86.91)] = -4.7038 \quad (5.10)$$

The previous velocity is not there for the first iteration, thus the value of v_{id} on right hand side of the equation is 0. c_1 and c_2 are taken as 2.05 in second and third term. The random number is taken as 0.263 from 5th row and 3rd column of Table 5.4. In actual execution these random numbers should be dynamically generated and should be different from each other in the second and third term. Since the individual best position and the current position of the particle are the same, thus the values for individual best term in the velocity calculation are also the same, i.e., 86.91. The last term gives the difference between the respective input variable values of the global best position and the current particle position.

Similarly, the remaining velocities for each variable and particle are calculated. The velocities for each particle in the first iteration are found and is represented in Table 5.5

Table 5.5
Velocity of Particles - Iteration 1

	Variable 1	Variable 2	Variable 3	Variable 4	Variable 5
Particle 1	−17.9029	−146.945	219.781	37.29145	−38.6636
Particle 2	0	0	0	0	0
Particle 3	−412.101	−8.54553	5.870366	−73.1169	55.92154
Particle 4	75.57413	25.94147	229.6472	21.35931	−9.93766
Particle 5	−680.304	−30.2667	−4.70385	−47.2952	99.4976
Particle 6	104.8392	−669.151	−42.8563	−2.51173	188.2581

It can be seen that the velocity of Particle 2 is 0. This is because it is the global best and it is also at its individual best position. In the first iteration, there will not be any contribution of individual best component in the calculation of velocity as the particles are already at their individual best position.

Now based on equation 5.8, let us update the position of the particles. This can be simply done by adding Table 5.2 to Table 5.5. After each update of particle position, it should be checked whether all of the variables in the swarm are within permissible range or not if not then, they should be properly rectified. The new upated position of the particles of the swarm is given in Table 5.6.

After updating the swarm, the objective function value for each particle shall be evaluated again. The results are shown in Table 5.7.

While comparing the data from Table 5.3 and 5.7, it can be observed that for particles 1, 3, 4 and 5 the Objective Function value has reduced in the second iteration. Thus the individual best of these particles shall be updated. For the Particle 6, the individual best shall be the same as that in iteration 1 and its individual best

Table 5.6
Swarm Positions - Iteration 1

	Variable 1	Variable 2	Variable 3	Variable 4	Variable 5
Particle 1	123.9783	−251.601	−177.825	249.0603	522.3422
Particle 2	−16.0475	−277.517	74.9442	247.9742	391.221
Particle 3	91.3609	194.3733	57.1011	283.4502	377.6338
Particle 4	−5.3758	−303.065	49.3234	98.8231	387.9588
Particle 5	−208.969	−182.907	82.2095	338.8905	−360.93
Particle 6	−451.232	−382.944	75.9312	247.3027	294.9124

Table 5.7
Objective Functions for the Swarm - Iteration 1

	Objective Function Value
Particle 1	112.2965
Particle 2	75.35377
Particle 3	69.0759
Particle 4	64.4123
Particle 5	83.2339
Particle 6	127.1777

position will not be changed. For Particle 2, since its position has not changed, its individual best also remains unchanged. It can also be observed that the global best position will also improve from 75.3538 to 64.6123, represented by the current position of Particle 4. The individual best matrix and global best position are represented in Table 5.8 and 5.9, respectively.

The update of Global Best Position marks the end of the first iteration. The calculation results of one more iteration is presented to observe the variation in velocity, individual best positions and global best position of the swarm. Now that the search process has started to roll, each particle shall have previous velocity and thus the first term cannot be 0 now, also some particles shall have values for the second term as well and Particle 2 which remained static during the first iteration with no velocity will also move.

Let the random values generated for this iteration be given by Table 5.10.

The velocities for iteration 2 generated from the random values given in Table 5.10 are tabularized in Table 5.11.

From Table 5.11, we can observe that now all the particles will have velocity. The particles which are at their individual best positions shall have that component as 0, particles which are at a position that is their individual best and also the global best, will have those components as 0, but still these particles shall have the inertia component and hence they will keep exploring the search space. The new updated

Table 5.8
Individual Best Position - Iteration 1

	Variable 1	Variable 2	Variable 3	Variable 4	Variable 5
Individual Best 1	123.9783	−251.601	−177.825	249.0603	522.3422
Individual Best 2	−16.0475	−277.517	74.94419	247.9742	391.221
Individual Best 3	91.36086	194.3733	57.10106	283.4502	377.6338
Individual Best 4	−5.3758	−303.065	49.32343	98.8231	387.9588
Individual Best 5	−208.969	−182.907	82.20951	338.8905	−360.93
Individual Best 6	−556.071	286.2077	118.7875	249.8144	106.6543

Table 5.9
Global Best Position - Iteration 1

	Variable 1	Variable 2	Variable 3	Variable 4	Variable 5
Global Best	−5.3758	−303.065	49.32343	98.8231	387.9588

Table 5.10
Random Numbers for Velocity Calculation - Iteration 2

	Variable 1	Variable 2	Variable 3	Variable 4	Variable 5
Particle 1	0.106653	0.084436	0.181847	0.54986	0.401808
Particle 2	0.961898	0.399783	0.263803	0.144955	0.075967
Particle 3	0.004634	0.25987	0.145539	0.853031	0.239916
Particle 4	0.77491	0.800068	0.136069	0.622055	0.123319
Particle 5	0.817303	0.431414	0.869292	0.350952	0.183908
Particle 6	0.868695	0.910648	0.579705	0.51325	0.239953

Table 5.11
Velocity of Particles - Iteration 2

	Variable 1	Variable 2	Variable 3	Variable 4	Variable 5
Particle 1	−33.6686	−113.617	221.9506	−96.2702	−108.881
Particle 2	15.34063	−15.2643	−10.1007	−32.3103	−0.37035
Particle 3	−301.092	−199.417	2.587858	−288.667	44.46876
Particle 4	55.09354	18.91133	167.4128	15.57093	−7.24456
Particle 5	−247.269	−99.5339	−46.1519	−160.389	278.3591
Particle 6	519.1425	531.5558	−17.1655	−113.792	103.0975

swarm given in Table 5.12, for the second iteration shall be generated by adding Tables 5.6 and 5.11. It is required to remind the users that for every update in swarm positions, the range of variables should be checked and the values of the variables should be kept within the allowable limits.

Table 5.12
Swarm Positions - Iteration 2

	Variable 1	Variable 2	Variable 3	Variable 4	Variable 5
Particle 1	90.3097	−365.218	44.12532	152.7901	413.4616
Particle 2	−0.70685	−292.781	64.84346	215.664	390.8506
Particle 3	−209.731	−5.04339	59.68892	−5.21692	422.1026
Particle 4	49.71774	−284.154	216.7363	114.394	380.7143
Particle 5	−456.238	−282.441	36.05765	178.5015	−82.5709
Particle 6	67.91099	148.6122	58.7657	133.5105	398.0099

The objective function values evaluated for the above updated swarm are given in Table 5.13. It can be observed that the objective function values are reducing for each iteration with some exceptions, however, this may not happen for all iterations.

Table 5.13
Objective Functions for the Swarm - Iteration 2

	Objective Function Value
Particle 1	85.1867
Particle 2	73.1457
Particle 3	56.8962
Particle 4	72.5798
Particle 5	82.9636
Particle 6	52.5815

The updated individual best and global matrices are given in Table 5.14 and 5.15.

Let us stop the calculation here as it is enough for the readers to further calculate the values required in the optimization process on their own.

It can be observed that the individual best position of each particle has improved. This may or may not happen depending on the random values. Also the variation in ω may affect the search process and the best positions may not change. As the iterations increase, the best positions shall seem to remain constant, this may be a signal to stop the search process. Many of the research papers shall plot the global best position variation over the iterations to show the improvement in search process.

The PSO program for above particles and objective function was executed for 200 times. The variation in the global best position is plotted for the execution in

Table 5.14
Individual Best Position - Iteration 2

	Variable 1	Variable 2	Variable 3	Variable 4	Variable 5
Individual Best 1	90.3097	−365.218	44.12532	152.7901	413.4616
Individual Best 2	−0.70685	−292.781	64.84346	215.664	390.8506
Individual Best 3	−209.731	−5.04339	59.68892	−5.21692	422.1026
Individual Best 4	−5.3758	−303.065	49.32343	98.8231	387.9588
Individual Best 5	−456.238	−282.441	36.05765	178.5015	−82.5709
Individual Best 6	67.91099	148.6122	58.7657	133.5105	398.0099

Table 5.15
Global Best Position - Iteration 2

	Variable 1	Variable 2	Variable 3	Variable 4	Variable 5
Global Best	67.91099	148.6122	58.7657	133.5105	398.0099

Fig 5.5. This method of showing the performance of an EOA is presented in many number of research papers and would be helpful for young researchers. The x-axis gives the number of iterations and the y-axis represents the objective function value.

As a matter of suggestion for those who plan to write the code of PSO on their own, keep the following in mind while writing the code: (i) generate the random numbers dynamically when required, and (ii) combine the objective function along

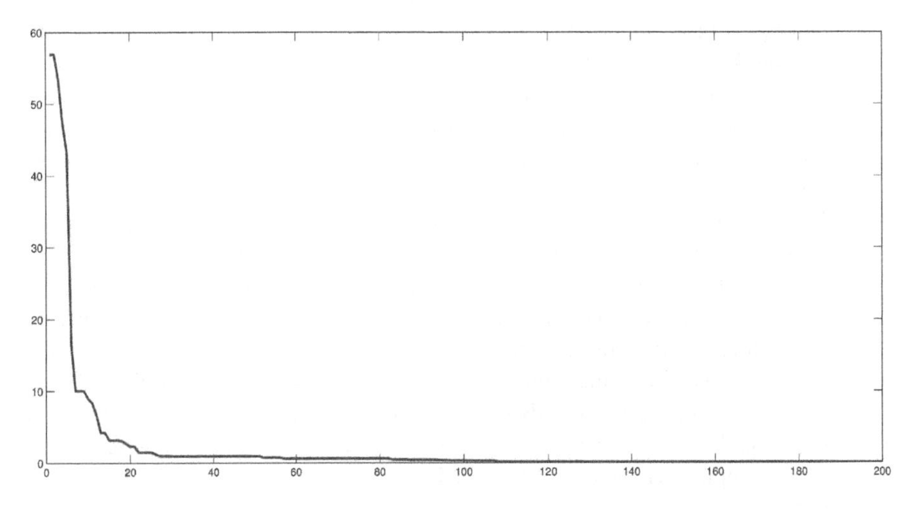

Figure 5.5: Global Best Variation over Iterations

with the swarm, individual best and global best positions for ease in writing the program.

5.8 VARIANTS AND HYBRID

There are a plenty of variants and hybrid versions of PSO available in the research domain. This section tries to cover as many variants and hybrids as possible, to let the reader get acquainted with various changes in PSO. Also it is more important to get the concept of a new variant or a hybrid version. Lastly, there may be multiple publications related to a variant or hybrid; however, we will discuss only one reference for one case. Some very specific case related PSO variants are not considered.

5.8.1 VARIANTS

1. **Bare Bones**: By definition, Bare Bones means bare necessity. In other words it tries to remove the unwanted parts of the process. In [108], the bare bones version of PSO is discussed. The velocity term is removed and a couple of Gaussian random numbers are included. The proposed equation for Bare Bones PSO is given in Eq 5.11.

$$x_{id}^{(k+1)} = N\left(\mu, \sigma^2\right) \tag{5.11}$$

where $x_{id}^{(k+1)}$ is the dth input variable value for i^{th} particle in $(k+1)^{th}$ iteration,

$$\mu = \left(p_{gd}^{(k)} + p_{id}^{(k)}\right)/2$$

and

$$\sigma^2 = \left|p_{gd}^{(k)} - p_{id}^{(k)}\right|$$

In above expressions, p_{id} is the personal best of the particle in consideration and p_{gd} is the global best of the swarm.

2. **Binary PSO**: The name of the variant itself explains the variation being introduced in PSO. In Binary PSO, the particle can have only 0 and 1 as their value. The immediate thought that strikes the mind should be, that if the values allowed within the particle are only 0 and 1, then how to evaluate the velocity? Well let's see how this is done.

   In this case, we will be refering [150] for understanding the basic concept of handling binary values within a particle and for its movement. It should be very clear that the particle (X_i) can have only 0 or 1 stored inside it. One can easily generate random particles having binary values only and create a swarm. For velocity, the sigmoid function is used. While dealing with velocity, the minimum (V_{min}) and maximum (V_{max}) value that can be stored in the velocity vector has to be decided. Also $V_{min} = -V_{max}$ is required. Another constraint mentioned in the reference is that the swarm size should be twice the dimension on input variables/particle size. To evaluate the initial velocity Eq 5.12 is used:

$$v_{id}^{0} = V_{\min} + (V_{\max} - V_{\min}) * \text{rand}() \tag{5.12}$$

After the iteration process starts, the velocity is evaluated as Eqs 5.13 and 5.14:

$$\Delta v_{id}^{k-1} = c_1 r_1 \left(pb_{id}^{k-1} - x_{id}^{k-1} \right) + c_2 r_2 \left(gb^{k-1} - x_{id}^{k-1} \right) \tag{5.13}$$

where pb_{id} is the individual best and gb^{k-1} is the global best, while other variables have their usual meanings.

$$v_{id}^{k} = h\left(v_{id}^{k-1} + \Delta v_{id}^{k-1} \right) \tag{5.14}$$

In which 'h' represents a piece wise linear function, such that if $v_{id} > V_{max}$, then $v_{id} = V_{max}$ and if $v_{id} < V_{min}$, then $v_{id} = V_{min}$, otherwise v_{id} retains its value. In this way the velocity vector shall have values in between V_{max} and V_{min} only. Now to change the position of the particle Eq 5.15 is used:

$$x_{id}^{k} = \begin{cases} 1, & \text{if} \quad U(0,1) < \text{sigmoid} \left(v_{id}^{k} \right) \\ 0, & \text{otherwise} \end{cases} \tag{5.15}$$

3. **Memetic PSO**: The memetic PSO presented in [116] will be discussed. Meme has been discussed in the Shuffled Frog Leaping Algorithm chapter. In the Memetic PSO, the authors of [116] introduce local search in each iteration. This neighborhood search can be applied according to the predefined schemata. In the proposed local search, it will not be performed on all particles but instead on global best and some selected individual bests or some of their combinations. The local search algorithm is performed as given in Algorithm 5.3:

Algorithm 5.3 Memetic PSO

```
while Local Search is applied on p_q^{t+1} do
   Obtain New Local Solution y
   if Obj(y) better than Obj(p_q^{t+1}) then
      p_q^{t+1} = y
   end if
end while
```

The local search is applied on particle represented by p_q^{t+1}. Solutions 'y' is obtained in this local search for which the objective function value is evaluated. If this objective function value of 'y' is better than p_q^{t+1}, then p_q^{t+1} is replaced by 'y'.

4. **Quantum PSO**: Another variant of PSO is presented in [94]. The particles instead of following classical Newtonian random motion rules, follows quantum mechanical rules. The QPSO in [94] shows how to create different quantum based wells centered around a global best vector. Then a Schrödinger equation is introduced to obtain the wave function. Also, probability density function of the particle position is found. A Monte Carlo like measurement method is then utilized to collapse the wave function into a desired region. This leads to

movement of particles in the search space. There are lots of equations in this variant and hence it is left to the reader to go through the reference.

5. **Fully Informed PSO**: In the fully informed PSO variant it is assumed that a particle is not only attracted towards the global best [92], but towards other particles too. These other particles in consideration form different topologies presented in [92]. Clerc's equation for velocity calculation is condensed and presented as in Eq 5.16.

$$\vec{v}_{t+1} = \chi\left(\vec{v}_t + \varphi\left(\vec{P}_m - \vec{X}_t\right)\right) \tag{5.16}$$

where $\chi = 0.729$, $\varphi = \varphi_1 + \varphi_2, \vec{X}_t$ is current position of particle and

$$\vec{P}_m = \left(\varphi_1 \cdot \vec{P}_i + \varphi_2 \cdot \vec{P}_g\right) / (\varphi_1 + \varphi_2)$$

In Fully Informed PSO, the calculations are modified as in Eq 5.17.

$$\vec{P}_m = \frac{\sum_{k \in \mathscr{N}} \mathscr{W}(k) \vec{\varphi}_k \otimes \vec{P}_k}{\sum_{k \in \mathscr{N}} \mathscr{W}(k) \vec{\varphi}_k} \tag{5.17}$$

where $\mathscr{N}$ is the set of neighbors, $\otimes$ is s point-wise vector multiplication, $\mathscr{W}$ describes an aspect of particle or it may be taken as a constant and

$$\vec{\varphi}_k = \vec{\mathrm{U}}\left[0, \frac{\varphi_{\max}}{|\mathscr{N}|}\right] \forall k \in \mathscr{N}$$

where U is a function which returns a vector whose positions are randomly generated following the uniform distribution between the two values within the brackets.

6. **Guaranteed Convergence PSO**: When all the particles converge on a position which is the best discovered until the current iteration then this phenomenon is termed as stagnation [162]. In guaranteed convergence PSO, the velocity of global best positioned particle is obtained with a different equation. The equation is given as Eq 5.18:

$$v_{\tau,j}(t+1) = -x_{\tau,j}(t) + \hat{y}_j(t) + w v_{\tau,j}(t) + \rho(t)\left(1 - 2r_{2,j}(t)\right) \tag{5.18}$$

where τ represents the global best position. The first two terms are used to reset the position to 'j', the third term is used for continuation of current search direction and the last term is used to generate a random sample. The random term ρ is used to determine the diameter of the search area around the global best position. The value of ρ is dependent on consecutive successes and failures.

7. **Unified PSO**: The Unified PSO [110] combines the effect of global best and local best on the movement of a particle. The following equations Eq 5.19, give the calculations for velocity based on global and local best:

$$\mathscr{G}_i(t+1) = \chi\left[V_i(t) + c_1 r_1 \left(P_i(t) - X_i(t)\right) + c_2 r_2 \left(P_g(t) - X_i(t)\right)\right] \tag{5.19a}$$

$$\mathcal{L}_i(t+1) = \chi \left[V_i(t) + c_1 r_1' \left(P_i(t) - X_i(t)\right) + c_2 r_2' \left(P_{g_i}(t) - X_i(t)\right)\right] \quad (5.19b)$$

where $\mathcal{G}_i$ is the normal velocity with global best, whereas $\mathcal{L}_i$ is the velocity based on local best. χ is the constriction factor, P_g is global best and $P_{g_i}(t)$ is the local best. In the next step the unified velocity is obtained as given in Eq 5.20:

$$\mathcal{U}_i(t+1) = (1-u)\mathcal{L}_i(t+1) + u\mathcal{G}_i(t+1), \quad u \in [0,1] \quad (5.20)$$

where ‘u’ is termed as unification factor. If ‘u’ is taken as 1 then the usual PSO is applied and if ‘u’ is taken as 0 then local PSO variant is applied, while $0 < u < 1$ denotes a composite PSO. In another addition to the variant, a random value is inserted with one of the terms in $\mathcal{U}_i$.

8. **Cooperative PSO**: A cooperative PSO is presented in [163]. In cooperative PSO, ‘n’ swarms are created for a ‘n’ dimensional optimization problem. Each swarm represents one input variable of the problem and tries to optimize that single input variable value. A representation is given in Fig 5.6.

Swarm 1

x_1	x_2	x_3	x_4	Obj Fun
-9	8	5	5	-665
5	8	5	5	189
2	8	5	5	72
0	8	5	5	64

Swarm 2

x_1	x_2	x_3	x_4	Obj Fun
-9	2	5	5	-725
-9	8	5	5	-665
-9	4	5	5	-713
-9	3	5	5	-720

Swarm 3

x_1	x_2	x_3	x_4	Obj Fun
-9	8	4	5	-666
-9	8	5	5	-665
-9	8	0	5	-670
-9	8	-8	5	-678

Swarm 4

x_1	x_2	x_3	x_4	Obj Fun
-9	8	5	40	-700
-9	8	5	10	-670
-9	8	5	5	-665
-9	8	5	7	-667

Figure 5.6: Cooperative PSO

From the figure, it can be observed that there are 4 input variables, thus 4 swarms are created. In swarm 1, only the first variable can vary, whereas in swarm 2 only the second variable can vary and so on. For swarm 1, the values of the remaining variables are contant and are obtained by taking the global best values of the other swarms as shown by arrows in the figure. The same method is applied to the other swarms for obtaining their constant input variable values. In other words, for swarm 2, the first input variable will get the value from the global best of swarm 1, the third input variable will get its value from the global best of swarm 3 and so on. The concept can be expanded to more number of variables.

9. **Comprehensive Learning PSO**: In Comprehensive Learning PSO [80], the basic concept of particle movement lies in the fundamental that following only

the global best is not a good strategy, instead a particle should follow the better particles. The equation for Comprehensive Learning PSO is given as Eq 5.21.

$$V_i^d = w * V_i^d + c^* rand_i^l * \left(pbest_{fi(d)}^d - X_i^d\right) \tag{5.21}$$

where $f_i = [f_i(1), f_i(2), \ldots, f_i(D)]$ defines the individual best positions of particles that the particle ‘i’ should follow. $pbest_{fi(d)}^d$ is the input variable value of any particle or even of the particle itself. Another factor P_C is used to determine if the particle will follow its own individual best or the individual best of another particle. A tournament selection method is applied to select $f_i(d)$ to be followed. f_i can be changed after some iterations in case the particle does not improve.

10. **Opposition-based PSO**: Opposition based PSO [166] is based on a principle of having an opposite population and accepting this opposite population if it is better than the existing population. There are many variations in this method as compared to the original PSO and has been made very complicated. There are two processes involved in the Opposition based PSO, first: finding of opposite position of a particle, given by Eq 5.22 and second: applying Cauchy mutation operator to mutate the g_{best} given by Eq 5.23.

$$OP_{i,j} = a_j^p + b_j^p - P_{i,j} \tag{5.22}$$

where $OP_{i,j}$ is the opposite particle of particle ‘i’ for input variable ‘j’, a_j^p and b_j^p are the minimum and maximum values for input variable ‘j’ in the current population and $P_{i,j}$ is the value of input variable ‘j’ for particle ‘i’.

$$\text{gbest'}(i) = \text{gbest}(i) + w(i)^* N(X_{\min}, X_{\max}) \tag{5.23}$$

where gbest is the global best position, ‘N’ is Cauchy distribution function to generate random number between $X_{\min}$ and $X_{\max}$ and w(i) is given as:

$$w(i) = \left(\sum_{j=1}^{PopSize} V[j][i]\right) / PopSize$$

where $V[j][i]$ is the velocity component of jth input variable for particle ‘i’, and PopSize is the population size. According to the algorithm a random number is generated to find if the opposite population is to be found or not. If the opposite population is to be found then, after evaluation of the new opposite swarm, the better solutions are chosen from the existing population and the opposite population to maintain the swarm size. In case the opposite population is not to be found then, the normal PSO is executed for all the particles in the population. Lastly the mutation is performed on the global best position at the end of the iteration.

11. **Spatial Extension PSO**: A radius is added to every particle in Spatial Extension PSO [71], to check if the particle is going to collide with another particle.

The collisions are to be avoided which will further stop the particles from clustering. Different strategies are employed for this phenomena and its after effects.

12. **MultiObjective PSO**: In a multiobjective problem, there are multiple objectives which are required to be optimized. Thus in Multi-Objective PSO, the PSO has to form a pareto front which represents the best outputs of the optimization technique. The solution may be dominated or non dominated as explained in earlier chapters. In [5], a global set of non dominated solutions is created and the solution nearest to the particle from this set is taken as global best. Remaining evaluations are more or less the same as the original PSO.
13. **Attractive Repulsive PSO**: In attractive repulsive PSO [132], there are two phases in the optimization process: attraction and repulsion. Attraction is similar to the traditional PSO. For repulsive phase, Eq 5.24 is used.

$$\vec{v}(t+1) = \omega \cdot \vec{v}(t) - \phi_1(\vec{p}(t) - \vec{x}(t)) - \phi_2(\vec{g}(t) - \vec{x}(t)) \tag{5.24}$$

The sign of the second and the third term are reversed to take the particle away from the individual and global best positions.

14. **MultiPhase PSO**: Multi-phase PSO [8] uses a different velocity equation as shown in Eq 5.25.

$$V_{i,n}(t+1) = C_v * V_{i,n}(t) + Cg^* G_i(t) + Cx^* X_{i,n}(t) \tag{5.25}$$

where C_v, C_g, and C_x are coefficients whose calculation depends on the phase and group of the particle. A given particle will be in a specific group and phase at a given time. C_g and C_x should have different signs. The phase of a particle may change if there is no improvement in its output. The particle velocities are also reinitialized randomly after a certain number of iterations. Also the particles do not move to a new position until the objective function value shows improvement.

15. **Fitness Distance Ratio based PSO**: In Fitness Distance Ratio based PSO [115], the particles are attracted towards particles having better objective function value in the neighborhood Eq 5.26 gives the expression to find the velocity:

$$V_{id}^{t+1} = \omega \times V_{id}^{t} + \Psi_1 \times (p_{id} - X_{id}) + \psi_2 \times \left(p_{gd} - X_{id}\right) + \psi_3 \times (p_{nd} - X_{id}) \tag{5.26}$$

The last term is an addition in velocity evaluation to include the neighborhood particle with better objective function value. Any particle ‘j’ from the neighborhood that maximizes the value of Eq 5.27 is selected as the neighborhood best particle.

$$\frac{\text{Fitness}(P_j) - \text{Fitness}(X_i)}{\left|P_{jd} - X_{id}\right|} \tag{5.27}$$

where ‘d’ is the respective input variable.

16. **Dynamic PSO**: The Dynamic PSO works on faster convergence of particles. In [10], the velocity evaluation of the particle is calculated as Eq 5.28:

$$X_i^{t+1} = X_i^t + [X^{*t} - X_i^t) \times (Obj(X^{*t}) - Obj(X_i^t)] \times SV_1 \\ +[X^{**t} - X_i^t) \times (Obj(X^{**t}) - Obj(X_i^t)] \times SV_2 + rnd \times signis() \times SV_3 \tag{5.28}$$

where SV's are used to limit the velocity term within limits, Obj is used to find objective function value and signis() function is used to generate 0 or 1 randomly. 'i' represents a specific particle number and 't' stands for the iteration number. Also 'X' represents the particle, 'X^*' represents the individual best and 'X^{**}' represents the global best. The equation has been modified in the above representation for better understanding.

Some of the other research work is listed below in Table 5.16:

Table 5.16
Variations in PSO

Sr. No.	Description	Reference
1	Change in Swarm Size, Neighborhood Search	[187]
2	Global Best/Leader is challenged by challengers who are young and have better leading powers	[21]
3	Imitates free electron movement (thermal and drift) in electric field	[148]
4	Overlaid Meta-Optimizer	[113]
5	Increase list of parameters for different scenarios	[112]
6	Inspired from Data parallelism	[19]

5.8.2 HYBRID PSO

It would be impossible to compile all the hybrid versions of PSO, thus a compilation of some references is presented for hybrid versions of PSO. It is suggested to the readers to understand the concept of the hybrid versions while referring to the literature and not the problem on which it is applied. The hybrid algorithms related to PSO are listed in Table 5.17.

A very in-depth analysis of PSO is presented in [165].

Table 5.17
Hybrid PSO

Sr. No.	Hybrid With	Reference
1	Genetic Algorithm	[45, 121, 140, 182]
2	Differential Evolution	[28, 70, 103, 109]
3	Ant Colony Optimization	[55, 57, 56, 144]
4	Chaotic	[83, 117, 118, 188]
5	Artificial Bee Colony	[79, 168]
6	Shuffled Frog Leaping Algorithm	[88, 106, 175, 189]
7	Simulated Annealing	[13, 43, 58, 102]
8	Tabu Search	[44, 134, 139, 184]
9	Gravitational Search Algorithm	[61, 96, 97, 122]
10	Sequential Quadratic Programming	[36, 95, 156, 164]
11	Immune Algorithm	[149, 188]
12	Binary PSO + Real Coded PSO	[154]
13	Taguchi	[9]

6 Artificial Bee Colony

6.1 INTRODUCTION

Bees have been known to work as a coherent group for a long time. These insects are really successful and cooperate amongst themselves in the collection of food. The bees are able to utilize their surroundings to the optimal level and gather large amount of nectar/food for themselves.

The Artificial Bee Colony (ABC) optimization method was introduced in [64] by Karaboga in 2005. The paper does not extensively explain the method and hence we refer to [65] by Karaboga and Akey as the main reference paper to understand the ABC algorithm.

ABC contains and mimics different types of bees involved in gathering nectar and their role towards the collective efforts of the food gathering process. The food itself is the main target and the best food source is the optimal solution that is to be found. The food source acts like a possible solution whereas the bees are used to modify the location of the food to get better results.

According to [65], ABC is based on reaction–diffusion equations developed as a model to depict the foraging behavior of bees by Tereshko et al. in [151, 152, 153]. The foraging behavior of bees in a colony is based on three factors: Food Sources, Employed Foragers and Unemployed Foragers. Foragers are the bees who go and search for food.

ABC is a robust optimization technique and has been widely used in various research areas by a number of researchers. The most attractive part of ABC, is the usage of simple equations which are less in number as well. In the following sections we will go through the details of the optimization method and understand the terms and processes involved in ABC. However, before we move to the next section it is necessary to mention here that like other optimization techniques, readers might be expecting that we would be generating random bees here. NO!!! We will be required to generate random food sources and bees will be the processes applied to improve the food sources. In fact Eq 6.1 is the bee.

As for the disadvantages associated with ABC, it must be mentioned here that the reference paper has a lot of unnecessary information which can lead to confusion.

In ABC, there are a couple of places where one can get confused, like:

- In case the best food source gets adandoned, then should it be replaced or not?
- When it is required that the best food source should be saved, is it the best food source yet or the best food source in the 'current' food source matrix?

It appears that the ABC method has many areas in which it can be improved or can be investigated. But before going through the list one should remember that these suggestions apply to the reference paper and it might have been already implemented

DOI: 10.1201/b22647-6

in other research articles. The literature review for similar articles have not been taken care of. Some of the improvement that can be thought of are:

- During the modification of candidate solutions, we will come across Eq 6.1, which suggests that only one input variable should vary while modifying the position of any solution. Why not all? and What should be the optimal set of input variables that can be modified?
- Different methods could be utilized for allocation of onlooker bees to food sources based on their respective probabilities.
- While replacing an abandoned food source by a random food source, should their objective function values be compared? This is a very important point, because there might be a situation in which a food source which is the best solution, but is thrown off because it gets abandoned.
- There is no mention of convergence in ABC, should it be included for a better output?
- In Eq 6.1, what would be the effect of ϕ on obtaining optimal solution? What should be its optimal range? Should its value be correlated with other parameters like 'j' and 'k' in the equation?
- Can the terminology be improved? Instead of using 'j' , 'k' and 'ϕ', could they be allotted some better terms? Why is no name given to the other food source which is used to modify a given food source? Best food source could also be given a name.
- Too much of randomness is there in ABC. In Eq 6.1, which is the main equation in the method, we have three parameters viz, 'j' , 'k' and 'ϕ, which are generated randomly.
- In Eq 6.1, the difference is obtained between the two input variables. Should it be difference only? Can it not be $\pm$, based on the relation between the food source getting modified and the food source modifying it?

Well the method offers a lot to explore and thus let us first explore the ABC method in the coming sections.

6.2 TERMINOLOGY

This section shall cover a number of terms used in ABC. These terms may or may not be relevant with respect to writing of the code for ABC. The terminology of ABC is:

1. **Food Source**: It acts as the main attraction for a honey bee during foraging. The food source has several properties which make it more attractive than the others such as nearness to the hive, energy content, taste, availability, etc. In ABC, these properties of the food source are represented by its objective function value. A food source is more attractive if it has a better objective function value.
2. **Employed Foragers**: Employed foragers are those bees which are associated with a particular food source. The employed forager bee exploits the food

source and shares its information with other bees in the hive. The information shared includes direction, distance, etc. of the food source from the hive. In ABC, the employed forager bee is actually a process applied to the food source to search for a better solution in its vicinity.

3. **Unemployed Foragers**: Bees that are not attached to a particular food source are termed unemployed foragers. Unemployed forager bees are classified as scouts and onlooker bees.
4. **Onlooker**: Onlooker bees are those bees which are not attached to a particular food source. They try to exploit a food source based on the information provided by the employed forager bees in the dancing area of the hive. The onlooker bees take clues from the information shared by the employed forager bees, then choose a food source and try to exploit it. In ABC, onlooker bees are supposed to perform the same operation as the employed forager bees on the food source, except that more onlooker bees will be attracted to the better food source. The role of the onlooker bees will be explained in more detail in later sections.
5. **Scouts**: Scouts are those bees that will search the surrounding area of the hive randomly. These bees do not take clues to search for a food source from any other bees. The scout bees account for around 5 - 10% of bees in the hive. In ABC, scout bees are used to generate random solutions. These bees and their role in the search process will be better understood in the following sections.
6. **Hive**: The hive is the place where all the bees live. The house of the bee colony is called the hive. The hive also houses the dancing area. The hive has no mathematical involvement in the ABC process.
7. **Dancing Area**: The employed forager bees are supposed to interact with the other bees in the dancing area. The employed forager bees dance in the dancing area to propagate the information related to the food sources to the onlooker bees. The onlooker bees watch the employed forager bees and select the food source they want to exploit. In ABC, a sequence of equations can be attributed to the job of performing the dancing area in the real hive.
8. **Nectar**: Nectar is the food available at the food source. The taste, quantity, energy content, etc. of the nectar shall decide the attractiveness of the food source. In other words, nectar is the objective function value of the food source.
9. **Waggle Dance**: The dance of the employed forager bees used to transmit the information related to the food source is termed as the waggle dance. The waggle dance shall be performed in the dancing area. The waggle dance shall show the attractiveness of the food source. The onlooker bees will decide the food source to be exploited based on the waggle dance. There are other dances also performed by the bees, but their discussion shall be out of the scope of the book. In ABC, the waggle dance is the process used to select the food sources which the onlooker bees will exploit.

6.3 FUNDAMENTAL CONCEPT

ABC was introduced in 2005 by Karaboga et al.[64]. Since this paper does not have much information about the process of ABC, another publication of Karaboga et al. from 2009 is taken as the main reference paper [65]. ABC tries to imitate the foraging behavior of the bees to generate the optimal process. The bees are a social animal and live in colonies. They cooperate with each other to gather food.

As mentioned earlier, ABC is based on foraging model of honeybees developed by Tereshko. This model is based on reaction–diffusion equations. The process of ABC is simple and is based on Food sources and Bees in the hive. The bees are classified as given in Fig 6.1.

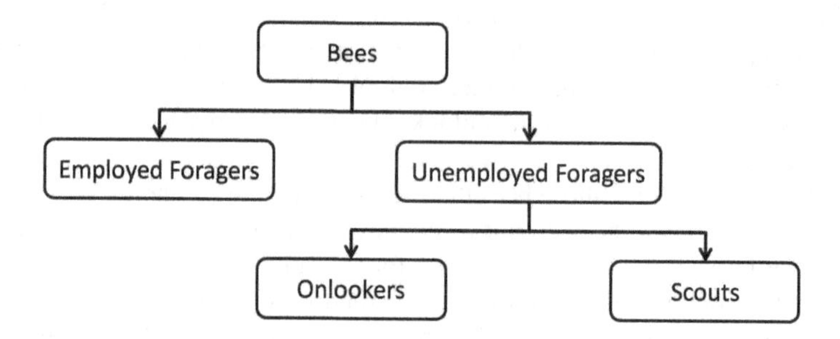

Figure 6.1: Classification of Bees

The bees will work on different food sources and improve the positions of the food sources. These food sources have different properties such as nearness to the hive, energy content, taste, etc., which decides its attractiveness. These properties are the input variables in ABC and the attractiveness of the food source is decided by its objective function value.

The main participants in the ABC process can be listed as: Food Sources, Employed Forager and Unemployed Foragers. The employed foragers are those bees which are linked to a particular food source. They visit the food source and try to manipulate its position to find better solutions. The unemployed forager are those bees which are not linked to any food source. These bees are further classified as onlooker and scouts. Onlooker bees may exploit food source whose position it acquires from the dancing area through the waggle dance of the employed forager. Onlooker bee does not randomly search for a food source. Scouts on the other hand are those bees which will randomly search for food sources and they account for 5–10% of the forager bee population.

The bees work together to collect food. For this cohesive collection of food, information needs to be shared amongst the bees. The bee hive has a dancing area dedicated to exchange of such information. The bees who want to share the information related to their known food sources and the properties of these food sources, perform a dance in this dancing area. The dance is termed waggle dance (there are some other terms also related to these dances of the bees). The dance passes the information about the food sources to the onlooker bees, who can use this information for further exploitation of the food source. Onlooker bees shall probably choose the most attractive food source, i.e., the solution with a better objective function value.

When a new unemployed forager bee starts its search, it has two options: to start as an onlooker bee or a scout. However, once the bee returns to the hive after visitng the food source, it becomes an employed forager. Now it has three options with itself: (1) it may abandon the food source, (2) it may attract more bees to the food source before returning to the food source or (3) it may exploit the food source without recruiting more bees. Also, all the bees will not start searching for food simultaneously. It should be noted here that the information in this paragraph is related to the actual bee colony; the ABC may have some differences in actual implementation. After understanding the actual foraging process of the honey bee, let's observe how it can be implemented in ABC.

In ABC, the food sources are possible candidate solutions for the optimization problem. The nectar to be collected from the food source is therefore equivalent to the objective function value of the possible solution. In ABC, the number of employed foragers or onlooker bees is the same as that of the number of solutions, i.e., the food sources.

Initially a population is generated randomly generated, which represents the positions of the food source and is denoted by 'SN'. Individual solution is represented by 'x_i' where $i \in (1,2,...SN)$. Also a single individual solution shall have all the properties of the food i.e., it will be a set of all input variables. The number of iteration for which the process of ABC shall be repeated is given by 'MCN' (Maximum Cycle Number) and its individual repetition is given by 'C' such that $C \in (1,2,....MCN)$. Each repetition cycle shall be applied for all the various types of bees.

The employed forager exploits the food source with which it is attached and tries to modify the position of the food source with which it is attached. If the new position of the food source has a better nectar availability i.e., objective function value, than the previous position, then the bee forgets the earlier position and remembers the new position. In other words, the food source i.e., the candidate solution, is modified. If the objective function value of the modified position is not better than the previous position, then the previous position is retained and the new position is not considered further. To modify the position of the food source Eq 6.1 is utilized.

$$v_{ij} = x_{ij} + \phi_{ij}\left(x_{ij} - x_{kj}\right) \tag{6.1}$$

In above equation, k ∈ (1,2,....SN) and j ∈ (1,2,...D). 'D' gives the number of input variables/parameters. Both 'k' and 'j' are randomly chosen indices, with a constraint of k ≠ i. Also ϕ_{ij} is a random number between -1 and 1.

After all the employed foragers have visited their food sources, they share the information related to the food sources with the onlooker bees. The food sources are selected by the onlooker bees based on the probability, which is related to the nectar available at these positions. The probability value associated with a food source is represented by p_i and is given by Eq 6.2. In Eq 6.2, fit_i, represents the objective function value of the food source 'i'. Even though the Eq 6.2 shows the probability of each food source to be selected by the onlooker bee, the exact method/process of selection, like the ones discussed in GA, is not given.

$$p_i = \frac{fit_i}{\sum_{n=1}^{SN} fit_n} \tag{6.2}$$

Similar to the employed forager, the onlooker bee will also modify the food source position and evaluate its nectar availability using Eq 6.1. If the new food source position has a better nectar availability, then the previous position is forgotten and the new food source position is remembered. In other words, the onlooker bee shall repeat the same process as that of the employed forager bee, except that one onlooker bee will not be mapped to one food source always, i.e., one food source may have more than one onlooker bee while the other food source may not have any onlooker bee.

In both the cases of employed foragers as well as the onlooker bees, if the new food source position is not better than the previous one than the previous food source position is retained and vice versa.

The question should now arise as to what happened to the SCOUTS??? Let's see further.

In every iteration of ABC, three processes are involved, first for employed forager, then for the onlooker bee and the last for the scout bees. In the third process, the number of scout bees is required to be determined. Scouts are required to generate random food sources when the other bees abandon the existing food source. The abandonment of a food source may happen when its food is no longer improving, i.e., the objective function value. A food source can be abandoned by the bees after a pre-specified number of iterations without improvement in the objective functon value. The scout bee is required to replace all these abandoned food sources by random food sources. In the reference paper, [65], it allows a maximum of one scout bee per iteration. It is also suggested that the number of employed forager bees and onlooker bees should be the same. Eq 6.3 provides the equation to generate a random food source by the scout bee, in which $x^j_{\min}$ and $x^j_{\max}$ gives the minimum and maximum limits of jth input variable, respectively.

$$x_i^j = x^j_{\min} + \text{rand}[0,1]\left(x^j_{\max} - x^j_{\min}\right) \tag{6.3}$$

To summarize, in ABC, intially the random solutions are generated and evaluated. The optimization process then goes into a looping cycle, in which each cycle has the following processes:

- Employed Foragers visit food sources, modify them, evaluate new positions and remember the better food position. (Eq 6.1)
- Probabilites are generated for each food source. (Eq 6.2)
- Onlooker Bees repeat the process of Employed Foragers, but based on probabilities related to the food sources. (Eq 6.1)
- Find abandoned food source, utilize scout bees to find new food sources randomly. (Eq 6.3)

This loop is repeated for a specified number of cycles. The process of ABC is explained objectively in the forthcoming sections.

6.4 ALGORITHM AND PSEUDOCODE

In this section, we will discuss the algorithm which should be used for ABC. It is one of the easiest optimization techniques to implement with only three equations to be implemented. Algorithm 6.1 gives the stepwise implementation of ABC. Any optimization technique would require some input data, which will change from situation to situation. The algorithm of ABC first generates 'SN' random food sources. Each food source will have 'D' values representing as many input variables. After random generation of food sources, its attractiveness is evaluated, i.e., the objective function value.

Once the food sources have been evaluated, each employed forager bee is attached to one individual food source. That is one employed forager for one food source. The bee is a procees applied to improve the position of the food source. A loop is used for a employed forager to improve the food source. The number of food sources and the employed foragers is the same. Thus the loop shall repeat 'SN' number of times, in which one employed forager modifies one food source.

Algorithm 6.1 Artificial Bee Colony

```
START
Input the required data
Generate random food sources.
Evaluate the food sources.
for 1 to MCN do
  for 1 to SN do
    Modify Food Sources by Employed Foragers (Algorithm 6.2)
  end for
  for 1 to SN do
    Generate probabilities for each Food Source (Eq 6.2)
  end for
  for 1 to SN do
    Modify Food Sources by Onlooker Bees (Algorithm 6.2)
  end for
  for 1 to SN do
    if Food Source is abandoned then
      Generate random food source through Scout Bee (Eq 6.3)
    end if
  end for
  Store the best food source.
end for
Display Results
STOP
```

The modification of a food source is shown in Algorithm 6.2. In the process of food modification, the bee first finds an alternative location to the existing food source in concern. A bee uses Eq 6.1 to modify the position of the food source. Once a new location of the food source has been determined, the limits of this position is checked. If the value of any of the input variables on the new position of the food source lies outside the allowable limits of that input variable, then the value of the variable need to be set to the limits. Next, the objective function value of the new food source position is found. If the objective function value of the new food source position is better than the previous one, than the old position of the food source is replaced by the new position. Otherwise the food source retains its old position.

Algorithm 6.2 Modified Food Sources

Find v_{ij} for the bee (Eq 6.1)
Check Limit of v_{ij}
Evaluate the new position of food source given by v_i
if New position v_i better than Old Position x_i **then**
  Replace Old position with new position
end if

Once the modification of all the food sources has been done by their respective employed forager bee, the probability associated with each one of them is found. Eq 6.2 is used to find the probabilities of each food source. The food source having better objective function value will have more chances of being chosen by an onlooker bee. After obtaining individual probabilities of each food source, the onlooker bees are allocated to each one of them. It may so happen that some food sources may have more than one onlooker bee associated with it while some other food sources may not attract a single onlooker bee. However, it is to be followed that the number of onlooker bee should be equal to 'SN'.

There are different ways to associate onlooker bees with food sources. Some of the methods have been discussed in the chapter on GA and SFLA. The implementation of this procees shall be shown in the Example section. Also, it is up to the reader to implement whatever onlloker bee distribution method he/she wishes to implement. In the Algorithm 6.1, the onlooker bees are associated with the food source through a loop. Each onlooker bee is supposed to visit the food source which has been allocated to it and modify the food source. The modification process of the food source remains the same as that of employed forager bee and is given in Algorithm 6.2.

Once the food source modification is completed by the onlooker bees, it is required to determine the number of scout bees required for the respective iteration. The number of scout bees shall depend on the number of abandoned food sources. A food source will be abandoned if its objective function value does not improve for a certain prespecified number of iterations. Thus we need to visit each food source and check whether it has been abandoned or not. In case the food source has been abandoned, a scout bee is employed to find a random food source position and replace the abandoned food source with the new one. Eq 6.3 is used to generate random food source position by the scout bee.

Once employed foragers, onlooker bees and scout bees have performed their allocated tasks, the best food source position is found and stored. This completes one iteration ABC optimization method. These iterations are repeated for 'MCN' number of times.

6.5 FLOWCHART

ABC Optimization technique is one of the most easy to implement. Fig 6.2 gives the main flowchart for ABC. It is assumed that the required input data shall be gathered beforehand. In the initial part of the process, we generate 'SN' number of random Food Sources within the available search space. Individual Food Source is denoted by 'x_i'. After generating random food sources, we need to evaluate its attractiveness, i.e., the objective function value. The looping will now start in ABC, up to 'MCN', i.e., Maximum Cycle Number. In each iteration/cycle/loop, first the employed forager

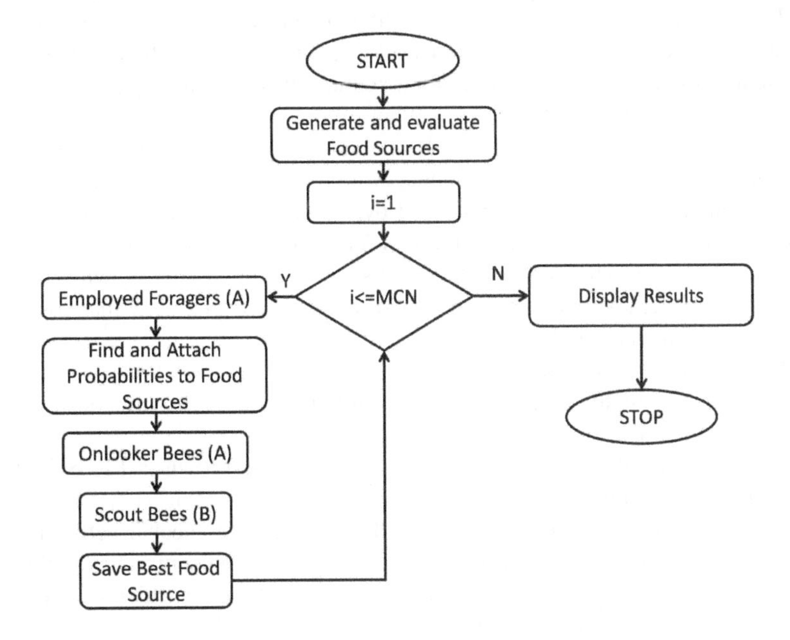

Figure 6.2: Flowchart for ABC

bees will visit the food source and modify its position. The employed forager bees will visit the food source with which it is attached. A bee basically represents a process to modify the food source, and hence in other words, each food source will be modified and compared with the existing food source. If the objective function value of the new/modified food source is better than the existing food source, then the existing food source shall be replaced by the new/modified food source. The flowchart of the modification process of each food source is given in Fig 6.3. The discussion on the modification procees will further be continued later in this section.

Let's move to the next step in the iteration, that is, the generation of probabilities and their attachment to the food sources. For each food source, a probability value is generated. The process of generating the probability can be taken from any of the selection methods employed in GA, like Roulette wheel, Rank Selection, etc. The method of allocating probabilties shall also be discussed in the Example Section. The probabilities are generated based on the objective function value of the food sources. Based on these probabilities, the onlooker bees will be allocated to the food sources. In simple words, by generating and allocating probabilities, we are attaching a number to each food source. This number shall be the number of times an onlooker bee shall visit that food source in the concerned particular iteration. That is, this number denotes the number of times it will be modified in that iteration.

Once the probabilities are given to each food source, the onlooker bees shall visit the food source and modify it. The modification process is the same as that of employed forager bees and is explained later. The process is shown as a flowchart in Fig 6.3. It should again be noted here that the number of onlooker bees visiting a food source shall depend upon the number allocated to it through the probability distribution and it may happen that some food sources may have multiple onlooker bees visiting it, while some food sources may have none.

After the onlooker bees have modified the food source positions, we need to find the food sources which have been abandoned. A food source shall be abandoned if its objective function value has not improved over a certain number of iterations. The scout bees shall be used to generate random food sources to replace the existing abandoned food sources. The implementation of scout bees in ABC is shown in Fig 6.4. The number of iterations required for a food souce to be declared as abandoned can be decided beforehand by the reader.

This finishes all the processes involved in a single iteration of ABC. One last step required to complete the iteration will be to store the best food source position, before moving on to the next iteration.

Let us now discuss the food source modification process shown in Fig 6.3. In the food source modification process, a modified food source position 'v_i' is first obtained by using Eq 6.1. Once 'v_i' is found, then we need to check whether the food source lies within the search space or not. In case the food source lies within the search space, then the input variables should be set to the limit which has been violated. The objective function value is then found for 'v_i'. If the new/modified food source position is better than the old/existing food source 'x_i', then the old food source is replaced with the new food source position. We then move on the next food source for modification.

For onlooker bees, the process remains the same, except that we need to take care as to which food source is required to be modified for a given iteration. For example, if the food source modification is to happen like this:

Food Source	Onlooker Bee
1	1
2	1
3	2
4	1
5	0
6	1

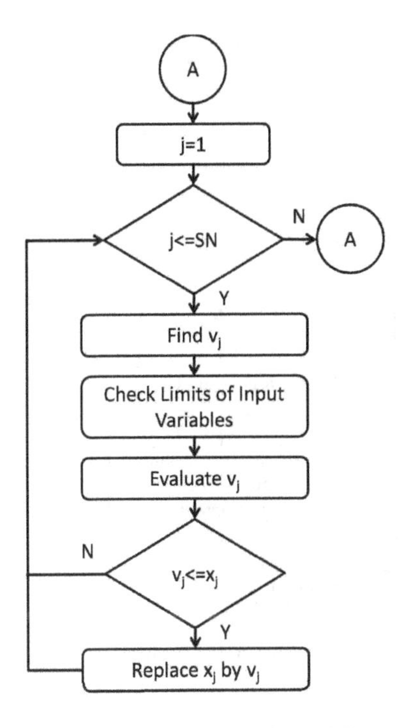

Figure 6.3: Modification of Food Sources

From the above listing it is understood that all the food sources shall be visited once, except Food Sources 3 and 5. The food source 3 shall be visited twice while the food source 5 wil not be visited at all. However, the total number of Onlooker bees i.e., 6, is the same as that of total number of food sources i.e., 6. Thus during iteration 1, food source 1 will be modified and during iteration 2, food source 2 will be modified. But the food source 3 will be modified twice during iterations 3 and 4. The remaining two iterations will modify food sources 4 and 6, respectively, without food source 5 being visited / modified.

Let us now move and examine the last flowchart of ABC, i.e., Fig 6.4, for scout bees.

For using scout bees, we need to find the abandoned food sources. Thus we loop through all the food sources to find those food sources which have not improved for a specified number of iterations. One needs to implement some mechanism to count such iterations is which a food source does not improve, for each and every food source.

If a food source is found to be abandoned, a random food source shall be generated using Eq 6.3. The random food source will be evaluated and it will replace the abandoned food source. It is usually assumed that the abandoned food source will represent a local optima and thus the food source is not getting modified.

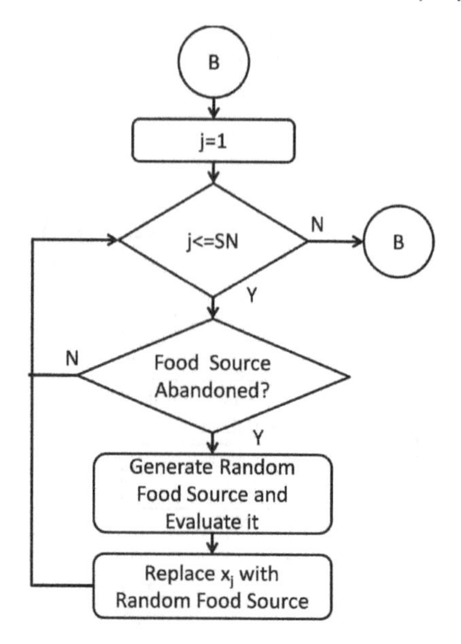

Figure 6.4: Scout Bees

6.6 EXAMPLE

The Griewank problem is solved for two iterations using ABC method in this section The calculations related to ABC are explained in detail in this section. In ABC, the fixed parameters are quite less. One such parameter is the number of iterations required to declare a food source as abandoned. The Griewank problem which is solved in this section, can be understood from Eqs 2.5 and 2.9.

The example will have the following: Food Sources - 6, Input Variables - 5 and number of iterations required to declare a food source as abandoned will be 5. But since we would be calculating the steps for two iterations only, no calculations will come up for abandoned sources / scout bees. Anyhow, the scout bee calculation only involves generation of a random solution, which would be done at the beginning of the example / program. The replacing of existing food source with the random food source can be achieved through an 'if' statement. Thus it is an easy process and is not required to be displayed here.

Let us observe the randomly generated food sources given in Table 6.1. The objective function value is clubbed with the food source here for better understanding. It is up to the reader to either keep the food sources separate from its objective function value or not.

As the number of food sources are taken as 6, the number of employed foragers and onlooker bees should also be 6. In the first part of the iteration, we employ the employed foragers to modify the food source positions. Eq 6.1 is used to modify the food source positions. In this example, we will be assuming that one input variable of the food source will be modified at a time and not all of them. Also, in this example,

Table 6.1
Food Sources in ABC

	Variable 1	Variable 2	Variable 3	Variable 4	Variable 5	Obj Fun
Food Source 1	141.8811	−104.656	−397.606	211.7688	561.0059	138.1064
Food Source 2	−16.0475	−277.517	74.9442	247.9742	391.221	75.3538
Food Source 3	503.4618	202.9188	51.2307	356.5671	321.7123	132.9249
Food Source 4	−80.9499	−329.007	−180.324	77.4638	397.8965	78.7628
Food Source 5	471.3352	−152.64	86.9134	386.1857	−460.428	154.5424
Food Source 6	−556.071	286.2077	118.7875	249.8144	106.6543	120.6237

we modify the food source immediately if a better food source is obtained by the employed forager bee. The reader must be aware while doing hand calculations about the value being considered. This would be shown in the coming calculations, through two tables. We need to remember that if a food source has been already updated during the current iteration then the updated values of input variables from that food source shall be considered.

Let's get down to business and check how the calculations are to be done in ABC. We now enter the iteration procees in ABC, and the first phase involves updating food sources by employed forager bees using Eq 6.1. For the first food source, i.e., i = 1, we chose another food source randomly, which comes out to be 6, i.e., k=6. Also the input variable to be modified is randomly chosen and it comes out to be 4, i.e., j =4. The value of ϕ was also randomly generated as 0.8049. Thus the Eq 6.1, can be written as:

$$\begin{aligned} v_{ij} &= x_{ij} + \phi_{ij}\left(x_{ij} - x_{kj}\right) \\ v_{14} &= x_{14} + \phi_{14}\left(x_{14} - x_{64}\right) \\ &= 211.7688 + 0.8049 * (211.7688 - 249.8144) \\ &= 211.7688 + (-30.6229) = 181.1459 \end{aligned}$$

The values which are not specified in the paragraph above are taken from Table 6.1. x_{14} means fourth input variable of the first food source and similarly x_{64} is the fourth input variable in the sixth food source. In other words the fourth input variable of the first food source is changed with the help of fourth variable of the sixth food source. v_{14} represents the fourth value of v_1. v_1 is similar to food source 1, except that its fourth variable is modified and thus it represents the modified position for food source 1. The next process is to find the objective function value of v_1, which is given in Table 6.2. It can be seen that only the fourth value of v_1 and food source 1 are different.

For v_1 represented in Table 6.2, the objective function value is evaluated to be 135.2946. The objective function value of Food Source 1 is 138.1064. This means that v_1 represents a better food source than food source 1. Thus we will replace the food source 1 by v_1. The food source with the change in Food Source 1 is given in

Table 6.2
v_1 in First Iteration

	Variable 1	Variable 2	Variable 3	Variable 4	Variable 5
v_1	141.8811	−104.656	−397.606	181.1459	561.0059

Table 6.3.

Table 6.3
Food Source 1 Modified by Employed Forager Bee

	Variable 1	Variable 2	Variable 3	Variable 4	Variable 5	Obj Fun
Food Source 1	141.8811	−104.656	−397.606	181.1459	561.0059	135.2946
Food Source 2	−16.0475	−277.517	74.9442	247.9742	391.221	75.3538
Food Source 3	503.4618	202.9188	51.2307	356.5671	321.7123	132.9249
Food Source 4	−80.9499	−329.007	−180.324	77.4638	397.8965	78.7628
Food Source 5	471.3352	−152.64	86.9134	386.1857	−460.428	154.5424
Food Source 6	−556.071	286.2077	118.7875	249.8144	106.6543	120.6237

So now with the Food Source 1 being modified by the employed forager bee, let's move to the Food Source 2. Let us remember one very important thing here, that after calculating the value of v_{ij}, we should first check whether it is within the allowable limits of the respective Input Variable or not. For this calculation, we have i = 2 and the random food source is 4, i.e., k = 4 and the input variable which will be changed is 5, i.e., j = 5. The value of ϕ is −0.9298. All these values have been randomly generated. Now to put them in Eq 6.1. All the other values in the calculation shall be taken from Table 6.3 as it represents the updated food source positions.

$$\begin{aligned} v_{ij} &= x_{ij} + \phi_{ij}\left(x_{ij} - x_{kj}\right) \\ v_{25} &= x_{25} + \phi_{25}\left(x_{25} - x_{45}\right) \\ &= 391.221 + (-0.9298) * (391.221 - 397.8965) \\ &= 391.221 + 6.2069 = 397.4279 \end{aligned}$$

The second food source is modified to v_2 which is given in Table 6.4. The objective function value of food position given by v_2 is found out to be 76.5844. The objective function value for v_2 is greater than the Food Source 2, thus the old position is retained and the food sources are still represented by Table 6.3.

The next food source to be visited by the employed forager bee is the third one. The random values generated are: Food Source = 2, i.e., k =2, Input Variable = 5 i.e., j =5 and ϕ = −0.9143. Also we understand now that i =3. The other values are to be

Table 6.4
v_2 in First Iteration

	Variable 1	Variable 2	Variable 3	Variable 4	Variable 5
v_2	−16.0475	−277.517	74.9442	247.9742	397.4279

taken from Table 6.3.

$$\begin{aligned} v_{ij} &= x_{ij} + \phi_{ij}\left(x_{ij} - x_{kj}\right) \\ v_{35} &= x_{35} + \phi_{35}\left(x_{35} - x_{25}\right) \\ &= 321.7123 + (-0.9143) * (321.7123 - 391.221) \\ &= 321.7123 + 63.5518 = 385.2641 \end{aligned}$$

The objective function value for the new position of Food Source 3 given by v_3 in Table 6.5 comes out to be 144.2699. Since v_3 does not have a better objective function value as compared to Food Source 3, the old position is retained and the food sources are still represented by Table 6.3.

Table 6.5
v_3 in First Iteration

	Variable 1	Variable 2	Variable 3	Variable 4	Variable 5
v_3	503.4618	202.9188	51.2307	356.5671	385.2641

The calculations for the remaining food sources is done in a similar manner. A summary of variables related to these calculations is presented in Table 6.6. The table contains the Food Source which is used in the calculations, the Input Variable which is being modified and ϕ.

The modified food sources are evaluated and replaced in the food sources if the new modified position is better than the existing food source position. The input variables 4, 5 and 1 are modified for Food Sources 4, 5 and 6, respectively. The new values of these input variables are found to be −122.438, −3.33356 and −224.314. Table 6.7 gives the values of 'v' for employed forager bees along with their evaluated objective function values.

It can be observed from Tables 6.1 and 6.7, that new modified positions of Food Sources 1, 5 and 6 are better than the existing food sources. Thus these food source are replaced by the new modified positions. The new food source matrix is represented by Table 6.8.

In the next step, we need to associate the onlooker bees to the food sources. Eq 6.2 is used to generate and allocate the probabilities with the food sources. One needs to

Table 6.6
Variables Used for Employed Foragers in Iteration 1

	Food Source (k)	Input Variable (j)	ϕ
Food Source 1	6	4	0.8049
Food Source 2	4	5	−0.9298
Food Source 3	2	5	−0.9143
Food Source 4	3	4	0.7162
Food Source 5	3	5	−0.5844
Food Source 6	4	1	−0.6983

Table 6.7
Modified Food Positions by Employed Forager Bee in Iteration 1

	Variable 1	Variable 2	Variable 3	Variable 4	Variable 5	Obj Fun
v_1	141.8811	−104.656	−397.606	181.1452	561.0059	135.2946
v_2	−16.0475	−277.517	74.94419	247.9742	397.4277	76.58444
v_3	503.4618	202.9188	51.23069	356.5671	385.2664	144.2699
v_4	−80.9499	−329.007	−180.324	−122.438	397.8965	81.16901
v_5	471.3352	−152.64	86.91336	386.1857	−3.33356	101.5441
v_6	−224.314	286.2077	118.7875	249.8144	106.6543	55.9908

Table 6.8
Food Sources after Employed Forager Bee

	Variable 1	Variable 2	Variable 3	Variable 4	Variable 5	Obj Fun
Food Source 1	141.8811	−104.656	−397.606	181.1452	561.0059	135.2946
Food Source 2	−16.0475	−277.517	74.94419	247.9742	391.221	75.35377
Food Source 3	503.4618	202.9188	51.23069	356.5671	321.7123	132.9249
Food Source 4	−80.9499	−329.007	−180.324	77.46379	397.8965	78.76276
Food Source 5	471.3352	−152.64	86.91336	386.1857	−3.33356	101.5441
Food Source 6	−224.314	286.2077	118.7875	249.8144	106.6543	55.9908

be careful while using this equation, as its application for maximization problem and that for minimization problem would be different. The current form of this equation is for maximization problem, we need to convert it into minimization problem. The generation of probabilities and their application is explained in the Roulette Wheel method of GA optimization method, and is therefore not explained here again (some other method can also be utilized). The number of Onlooker Bees associated with the food sources is given in Table 6.9, after implementing Roulette Wheel method of allocation. The total number of Onlooker Bees are 6, however, Food Source 3 does not get a single Onlooker Bee and Food Source 6 gets two Onlooker Bees. Thus Food Source 6 will be modified, two times, using Eq 6.1.

Table 6.9
Onlooker Bees Associated with Food Sources in Iteration 1

Food Source	Number of Onlooker Bees
1	1
2	1
3	0
4	1
5	1
6	2

The food source used for modifying, the input variable and the value of ϕ which are randomly generated for the Onlooker Bees is presented in Table 6.10.

Table 6.10
Variables Used for Onlooker Bees in Iteration 1

	Food Source (k)	Input Variable (j)	ϕ
Food Source 1	6	4	−0.5155
Food Source 2	5	3	−0.311
Food Source 4	2	1	0.8057
Food Source 5	4	2	−0.9004
Food Source 6	1	3	0.2369
Food Source 6	5	4	0.6263

Before we move ahead, the following points should be kept in mind: (i) the food source matrix shall change if the modified food source is better than the existing respective food source and (ii) the Food Source 3 has no row in Table 6.10 whereas, Food Source 6 has two. The remaining values shall be used from Table 6.8.

Let us perform some sample calculations. Equation to be used for Onlooker bees remains the same as Eq 6.1. From Table 6.10, we can observe that the Food Source 1 is modified at Input Variable 4 with the help of Food Source 6. And the value ϕ is

−0.5155. We therefore have, i = 1, j = 4 and k = 6.

$$\begin{aligned} v_{ij} &= x_{ij} + \phi_{ij}\left(x_{ij} - x_{kj}\right) \\ v_{14} &= x_{14} + \phi_{14}\left(x_{14} - x_{64}\right) \\ &= 181.1452 + (-0.5155) * (181.1452 - 249.8144) \\ &= 181.1452 + 35.399 = 216.5442 \end{aligned}$$

Replacing the fourth variable in Food Source 1 and obtaining the objective function value, we get 138.6828. The old position is better than the new position and hence the old position is retained for Food Source 1.

In Food Source 2, the value of i = 2, j = 3, k = 5 and ϕ = −0.311, as can be seen from Table 6.10. The value of v_{23} comes out to be 78.6661 which when replaced in the Food Source 2, give the objective function value of 75.5007. When compared to the objective function value of Food Source 2 in Table 6.8, it can be observed that the new objective function value is not better. Thus the Food Sources in Table 6.8 shall be maintained as it is.

Food Source 3 shall not be modified as per Table 6.9. For Food Source 4, the parameters to be used for modification are: i = 4, j = 1, k = 2 and ϕ = 0.8057. If the position of Food Source 2 was improved during the Onlooker Bees phase then we would be required to use the new position of Food Source 2. Right now we will be using the values from Table 6.8. The value of v_{41} is found to be −133.2442 and that of the objective function with this change in Input Variable value is 81.6564. Food Source 4 like other food sources is also not improved by the Onlooker Bee as its previous objective function value is 78.7628.

Food Source 5 has the following parameters: i = 5, j = 2, k =4 and ϕ = −0.9004. The value of v_{52} is found out to be −311.4486 which makes the objective function value of Food Source 5 to be 119.9740. Hmmmm!!!! The random numbers are not going in our favour. Food Source 5 had an objective function value of 101.5441, and thus we could not find a better position.

The Onlooker Bees have not been able to improve the food source positions until now. Food Source 6 has two Onlooker Bees and let's see if any one of the bee is able to improve its position. For the first Onlooker Bee the parameters are: i = 6, j = 3, k = 1 and ϕ = 0.2369. The value of v_{63} comes out to be 241.1124 and the respective objective function value for Food Source 6 is found to be 67.0111, which unfortunately is not better than the existing objective function value. In case, the Onlooker Bee was able to improve the position of Food Source 6, then in the next calculation we would have used this new position.

Let's try our luck with the last calculation of this iteration and see if the Food Source 6 is improved. Before we move ahead, just a reminder that we are still refering to Table 6.8 for the variable values. In this last calculation, i = 6, j = 4, k = 5 and ϕ = 0.6263. The value of v_{64} comes out to be 164.4112 which gives the objective function value of Food Source 6 as 47.1391. Finally, we got a better objective function value and Food Source 6 is modified. The new food source matrix is given in Table 6.11.

One iteration of ABC has three phases: Employed Forager Bee, Onlooker Bee and Scout Bee. In the last phase, i.e., Scout Bee phase, there will be no scout bees

Table 6.11
Food Sources after Onlooker Bee Modifications in Iteration 1

	Variable 1	Variable 2	Variable 3	Variable 4	Variable 5	Obj Fun
Food Source 1	141.8811	−104.656	−397.606	181.1452	561.0059	135.2946
Food Source 2	−16.0475	−277.517	74.94419	247.9742	391.221	75.35377
Food Source 3	503.4618	202.9188	51.23069	356.5671	321.7123	132.9249
Food Source 4	−80.9499	−329.007	−180.324	77.46379	397.8965	78.76276
Food Source 5	471.3352	−152.64	86.91336	386.1857	−3.33356	101.5441
Food Source 6	−224.314	286.2077	118.7875	164.4112	106.6543	47.1391

for this iteration. It was decided in the beginning of the section that the food source would be abandoned if the respective food source does not show improvement for 5 consecutive iterations. Infact we will not be having the scout bees in the next iteration also, and up to the fifth iteration. So we will be skipping this part of ABC method.

Table 6.11 gives the food sources to be used for the next iteration. Now that we understand the evaluation of v_{ij}, it may be not required to show every calculation and the random values of j, k and ϕ are just specified. The reader is supposed to calculate the results with these specified random values and verify the same. For the employed forager bee phase of ABC in the second iteration, the random values of j, k and ϕ were generated as given in Table 6.12. It should be kept in mind that the Food Source number and 'k' should not be the same.

Table 6.12
Variables Used for Employed Foragers in Iteration 2

	Food Source (k)	Input Variable (j)	ϕ
Food Source 1	3	3	−0.2926
Food Source 2	5	4	0.4479
Food Source 3	4	4	0.6748
Food Source 4	2	3	−0.9195
Food Source 5	3	3	0.5524
Food Source 6	5	2	−0.0119

Instead of calculating each and every value, a tabulated summary of variables evaluated and the objective function values obtained for their respective food sources is presented in Table 6.13. It is suggested that the reader calculates all the values to understand the process properly.

The reader should keep in mind that the food source matrix should be updated whenever the modified food source is better than the old food source position. In the above calculations, we find that food sources 1, 2, 4 and 6 get a better modified food source position. Thus these food sources are updated in the original food

Table 6.13
v_{ij} and Objective Function Values for Employed Foragers in Iteration 2

	v_{ij}	Obj. Fun. Value
Food Source 1	−266.265	113.4965
Food Source 2	186.0624	68.65682
Food Source 3	544.8994	175.3733
Food Source 4	54.39204	71.68162
Food Source 5	106.6236	102.4949
Food Source 6	280.9792	46.58402

source matrix. Also, from Table 6.12, it can be observed that while food sources 4, 5 and 6 are getting updated they use the food sources 2, 3 and 5, respectively. While evaluating, Food Source 4, we need to use updated Food Source 2. If one does not update immediately in the above case, still the output shall remain the same as the Input Variable getting modified are different. Thus one needs to be careful while evaluating the results with hand calculations.

The food source matrix after the employed forager bee phase of iteration 2 is given in Table 6.14. Note that the food sources and their respective input variables are modified as per Table 6.13.

Table 6.14
Food Sources after Employed Forager Bee modifications in Iteration 2

	Variable 1	Variable 2	Variable 3	Variable 4	Variable 5	Obj Fun
Food Source 1	141.8811	−104.656	−266.265	181.1452	561.0059	113.4965
Food Source 2	−16.0475	−277.517	74.94419	186.0624	391.221	68.65682
Food Source 3	503.4618	202.9188	51.23069	356.5671	321.7123	132.9249
Food Source 4	−80.9499	−329.007	54.39204	77.46379	397.8965	71.68162
Food Source 5	471.3352	−152.64	86.91336	386.1857	−3.33356	101.5441
Food Source 6	−224.314	280.9792	118.7875	164.4112	106.6543	46.58402

Now for the execution of Onlooker Bees phase. The Onlooker Bees are generated as in Iteration 1 and the distribution again comes up similar to the previous iteration as given in Table 6.9. It can be observed that if we use Roulette Wheel method, Food Source 3 has only 9% share approximately and hence it again does not attract an Onlooker Bee. It can further be understood that if a Food Source does not improve, then, it also stands to lose the Onlooker Bees which would further deteriorate its chances of improving. This would lead to a situation where finally such a food source shall be replaced by a random food source generated by the Scout Bee.

Also the Food Source 6 has aproximately 27% share and gets two Onlooker Bees.

Let's tabularize the values of 'j', 'k' and 'ϕ' used for the calculation of Onlooker

Bees for Iteration 2. Table 6.15 gives the values of these random variables used in Onlooker Bee phase along with the food sources where they are applied.

Table 6.15
Variables Used for Onlooker Bees in Iteration 2

	Food Source (k)	Input Variable (j)	ϕ
Food Source 1	4	5	−0.9186
Food Source 2	4	2	0.7014
Food Source 4	5	2	−0.8585
Food Source 5	3	2	0.4978
Food Source 6	4	3	0.2967
Food Source 6	5	3	−0.0994

Table 6.16 gives the modified input variable values and the objective function value for the respective food sources.

Table 6.16
v_{ij} and Objective Function Values for Onlooker Bees in Iteration 2

	v_{ij}	Obj. Func. Value
Food Source 1	411.1764	76.95205
Food Source 2	−241.401	64.02496
Food Source 4	−177.591	52.50649
Food Source 5	−329.649	122.8898
Food Source 6	137.8924	47.59479
Food Source 6	115.6177	46.14548

In the above calculations, the Food Source 6 is modified twice. The Input Variable getting modified is the same in both the cases. However, the first Onlooker Bee does not get a better objective function value and hence the food source remains the same. In case, the first Onlooker Bee had obtained a better objective function value, then the Food Source 6 would have been modified and we would be required to use the modified food source for the second Onlooker Bee.

The final food source matrix at the end of Iteration 2 is given in Table 6.17. The food sources 1, 2, 4 and 6 are modified in the Onlooker Bee phase of Iteration 2.

The ABC method was executed for the above example for 200 iterations. The variation of the best food source during these iteration is presented in Fig 6.5. The minimum objective function value at the end of all the iterations comes out to be 0.2028.

Table 6.17
Food Sources after Iteration 2

	Variable 1	Variable 2	Variable 3	Variable 4	Variable 5	Obj Fun
Food Source 1	141.8811	−104.656	−266.265	181.1452	411.1764	76.95205
Food Source 2	−16.0475	−241.401	74.94419	186.0624	391.221	64.02496
Food Source 3	503.4618	202.9188	51.23069	356.5671	321.7123	132.9249
Food Source 4	−80.9499	−177.591	54.39204	77.46379	397.8965	52.50649
Food Source 5	471.3352	−152.64	86.91336	386.1857	−3.33356	101.5441
Food Source 6	−224.314	280.9792	115.6177	164.4112	106.6543	46.14548

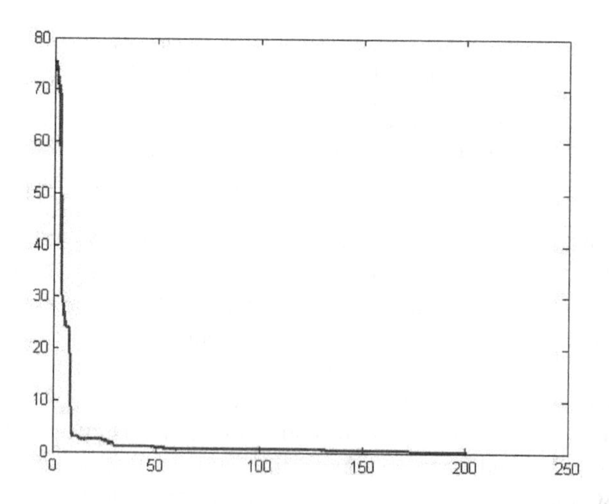

Figure 6.5: Variation of Best Food Source through Iterations in ABC

7 Shuffled Frog Leaping Algorithm

7.1 INTRODUCTION

Shuffled Frog Leaping Algorithm (SFLA) was introduced in 2006 by Muzaffar Eusuff et al. [41]. The SFLA is based on a swamp containing frogs. The method assumes that the frogs can be divided into different memeplexes. The frogs are carriers of memes which is utilized for optimization of a given system. The memeplexes are used to carry out the local searches. SFLA performs the local search a number of times within a memeplex and then rearranges the frogs amongst the memeplexes so that the global exploration can also be achieved. In case of saturation in the search process, random frogs are generated and introduced into the population.

SFLA is a memetic algorithm, which utilizes population based search for solving optimization problems. Memetic Algorithms are found to perform better than other meta-heuristic methods [41]. Meme is pronounced as 'meem' and it means a pattern of information which replicates itself and is contagious in nature. This pattern may contain information related to songs, fashion, phrases, technical, etc. The memes are imitated and spread from brain to brain. Memes also have a property, that, they can be improved by the person or animal holding it. Memes actually does not have anything to do with frogs, because frogs are only considered as hosts.

SFLA is a complicated algorithm. Even though the formulae used are easy the programmer will have to put in a lot of efforts to realise the actual SFLA method through coding. Some of the other drawbacks or area of research, found in the optimization technique are:

- It has not been discussed as to how to optimize the number of memeplexes 'm', number of frogs in each memeplex 'n' and also the number of frogs in each submemeplex 'q'.
- Similar to the above point, it is also not discussed as to how to optimize the step size.
- Only one frog is improved after all the classifications. Methods can be introduced for increasing this number.
- Chances given to the worst frog to improve is very constrained and limited.
- After generating the random frog to replace the worst frog in the submemeplex, the objective function values are not compared. There are chances that the newly generated random frog may not have a better objective function than the frog being discarded.
- The step size appears to be discrete in nature. This is further explained in the Example Section.

DOI: 10.1201/b22647-7

• There is no mention of selecting distinct frogs in the submemeplex. However, it is assumed so in this chapter.

Let us now explore the world of frogs!

7.2 TERMINOLOGY

This section shall be used to present the different terms used in SFLA. It will help the reader to get acquainted with the process. It should also be noted that some of the terms have been introduced as they were not defined in [41], where SFLA was introduced.

1. **Meme**: Meme is an information pattern, which parasitically infects human / animal minds and is contagious. All these properties makes the propagation of a meme, possible.
2. **Memotype**: It is the actual information in a meme. It is similar to a chromosome of a gene.
3. **Frog**: It is a probable solution. Similar to particle in PSO and a chromosome in GA.
4. **Swamp**: It is the feasible search space, in which the optimal solution is to be searched.
5. **Leap**: Step size with which the memetic vector / solution changes to take the frog to a new position.
6. **Stone**: Stones represents discrete locations in a search space. The frogs can take leaps onto the stones.
7. **Food**: Food is available on top of the stones. The food can be considered to be analogous with the objective function value.
8. **Memeplex**: The frogs are divided into different parallel groups / communities before the start of search process or each iteration. These groups are termed as memeplex.
9. **Submemeplex**: A subset of a memeplex is termed as a submemeplex.
10. **Memetic Evolution**: In a memeplex, a frog is infected with a better meme from another frog, and moves towards the optimal solution/position, termed as memetic evolution.
11. **Step Size**: The input variables are varied for each frog from one iteration to the other. The combination of variations for all input variables of a frog is termed as a Step Size. It is similar to velocity in PSO.
12. **Censorship**: Censorship is a move through which the infeasible frog or a frog which is not better than the old one, is stopped from spreading the defective meme and is replaced by a randomly generated frog.

7.3 FUNDAMENTAL CONCEPT

SFLA is an optimization technique which considers frogs as hosts to carry memes. In a memetic algorithm, the hosts are required to pass on the memes which contain the memotypes. The host has the choice to change the information contained in the meme. The meme can be spread selectively and rapidly, as it need not wait for the next iteration and can be passed from one host to the other as a part of increased communicability.

SFLA is a combination of deterministic and random approaches and can thus perform an informed heuristic search. In SFLA, the host / individual is not so important as compared to the information that it carries through the meme. The frog is used as a host to carry one meme, i.e., one possible solution to the optimization problem.

A group of frogs is considered to inhabitate the swamp. These frogs are required to find the position having optimal food allocation, which is discretely placed on stones. The frogs are required to leap and gather on stones having maximum available food. While moving, the frogs are assumed to communicate and improve their own memes and that of others.

In the initial step of SFLA, frogs are generated randomly. The randomly generated population of frogs is assumed to cover the whole swamp, i.e., search space, referred as 'Ω' . An interaction between different frogs will lead to memetic evolution, in which one frog is infected with the information / ideas of other frogs. This will lead to improvement in the performance of the frog while moving towards the optimal solution.

The total number of frogs 'F' in the swamp is given by 'm*n', where 'm' represents the number of memeplexes and 'n' represents the number of frogs in one memeplex. This is very important, as we need to decide the number of memeplexes and number of frogs in each memeplex before starting the optimization process. The frogs are randomly generated similar to the chromosomes, particles, etc. in other optimization techniques. Each frog should contain information related to each input variable, within the specified limits of the input variables. A frog shall represent one candidate / possible solution.

After randomly generating the frogs in the swamp, the objective function value of each frog is evaluated. The frogs are then sorted based on decreasing objective function values. The sorted list of frogs is stored as X = U(i), f (i), i = 1, . . . , F. One record in the sorted matrix 'X' will contain the possible solution, i.e., a frog given by 'U(i)', along with its objective function value given by 'f(i)'. Also the best frog shall be placed on the top, such that X(1), will give the frog with the best objective function value in the current population. The best frog position, i.e., X(1) is stored in 'P_x', as well.

The sorted matrix 'X' is now used to partition the frogs into 'm' memeplexes. The process of partitioning is explained in Fig 7.1.

Figure 7.1 shows how 12 frogs, which are stored in a sorted order of their objective function values, are distributed amongst 3 memeplexes. In the example given in Fig 7.1, n = 4 and m = 3. The distribution of frogs into memeplexes is given through Eq 7.1.

$$\begin{aligned} Y^k = & \left[U(j)^k, f(j)^k | U(j)^k = U(k+m(j-1)), f(j)^k \right. \\ & \left. = f(k+m(j-1)), \quad j=1,\ldots,n\right], \quad k=1,\ldots,m \end{aligned} \tag{7.1}$$

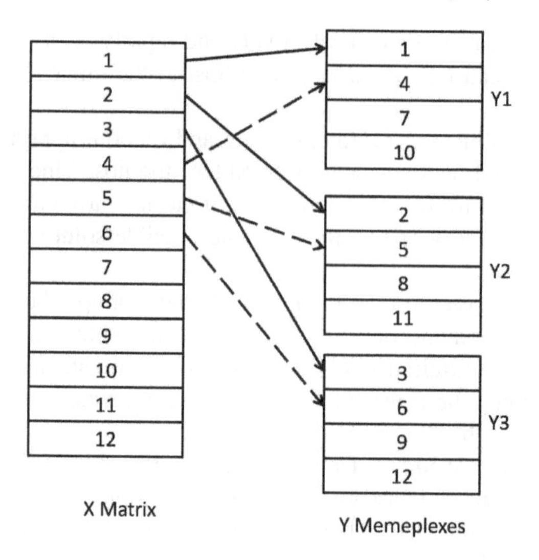

Figure 7.1: Distribution of Frogs into Memeplexes

In Eq 7.1,

$Y^k \rightarrow$ A single memeplex 'k', such that k $\in$ m

Let us put some random numbers in Eq 7.1 and see how it works,

Assume the values as: k = 1, j = 1, also from the figure we have, n = 12 and m = 3. When k = 1, we are considering the first memeplex. Thus we get:

$$Y^1 = [U(k+m(j-1)), f(k+m(j-1))]$$
$$= [U(1+3(1-1), f(1+3(1-1)] = [U(1), f(1)]$$

This means that the first element from 'X' goes to memeplex 1.

Let's try another one, where k = 1 and j = 2. This means, we are trying to find the second element of memeplex 1.

$$Y^1 = [U(1+3(2-1), f(1+3(2-1)] = [U(4), f(4)]$$

Similarly, let's try for the second element of third memeplex, i.e., k = 3 and j = 2.

$$Y^3 = [U(3+3(2-1), f(3+3(2-1)] = [U(6), f(6)]$$

The above can be verified from the Fig 7.1. Thus we can distribute all the frogs into different memeplexes in a similar fashion.

After dividing the frogs into different memeplexes, individual memeplexes are evolved. This evolving within the memeplexes is done for a specified number of times given by 'iN'. The frogs can move towards the frog with the best meme within the memeplex, however, this can lead to all the frogs getting positioned around the local minima, i.e., the best frog in the memeplex. To avoid this, the first step before memeplex evolution is to create a submemeplex. While creating the submemeplex, more weightage is given to the frog with better objective function value and lesser weightage to the frog with lower objective function value. The weights are assigned to the frogs through triangular probability distribution. Since the memeplexes are already sorted, the frog with the best objective function value is at the top and that with the worst objective function value in the memeplex is at the bottom. The triangular probability distribution method uses the formula given in Eq 7.2.

$$p_j = \frac{2(n+1-j)}{n(n+1)} \qquad j = 1,\ldots,n. \tag{7.2}$$

Thus for a memeplex having 4 frogs, i.e., n = 4, the value of p_j for the best frog will be $p_j = 2^*(4+1-1)/(4^*(4+1)) = 8/20 = 0.4$. Similarly, for the worst frog, $p_j = 2^*(4+1-4)/(4^*(4+1)) = 2/20 = 0.1$. The method is termed as triangular probability distribution because the probabilities decrease as we move from the best frog to the worst frog, as shown in Fig 7.2. In programming, a random number is generated between 0 and 1, and the frog to be selected for submemeplex shall depend on the region in which the random number lies. The individual probability range of each frog is also represented in Fig 7.2. It can be seen that Frog 1, i.e., the best frog in the memeplex shall be selected in the submemeplex if the random number lies between 0 and 0.4. Second frog will be selected if the random number lies between 0.4 and 0.7 and so on. It can be observed that the chances of the worst frog getting selected in the submemeplex are quite less as compared to the best frog.

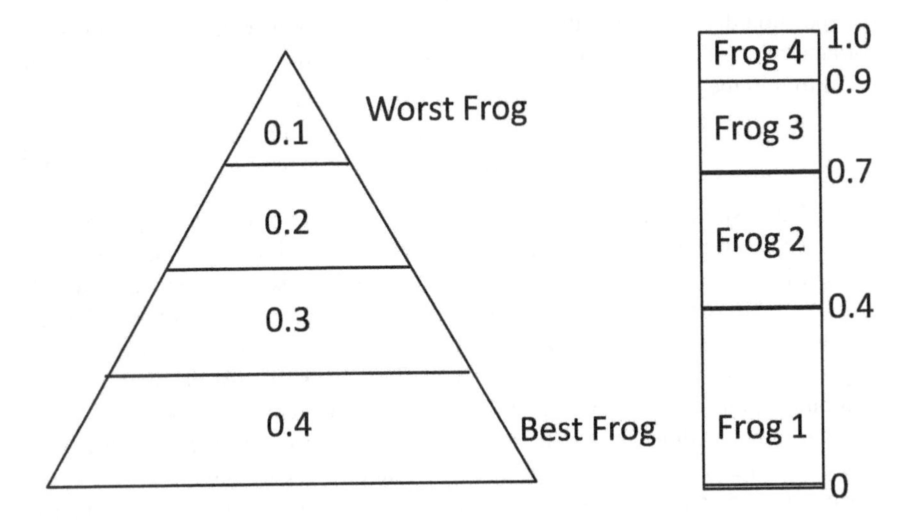

Figure 7.2: Triangular Probability Distribution

The number of frogs selected in the submemeplex is given by ‘q’ and the submemeplex is represented by ‘Z’. The ‘Z’ matrix is again sorted according to the objective function values to obtain the best frog represented by ‘P_B’ and the worst frog represented by 'P_W' in the submemeplex. The position of the worst frog is then improved using Eq 7.3.

$$\begin{aligned} S &= \min\{\text{int}[\text{rand}(0,1)(P_B - P_W)], S_{\max}\} \quad \text{for a positive step} \\ &= \max\{\text{int}[\text{rand}(0,1)(P_B - P_W)], -S_{\max}\} \quad \text{for a negative step} \end{aligned} \tag{7.3}$$

Eq 7.3 gives the step size for the worst frog to move to a better position with the help of the best frog within the submemeplex. It needs to be clarified here that the step size (by terminology) gives the vector which will represent the movement of the frog in the search space. However, Eq 7.3 has to be executed for each input variable of the worst frog. Also, the maximum step size has to be defined for each input variable if these variables are having different characteristics. It will be better clarified once the reader goes through the Example section. The terms used in the equation are:

rand(0,1) $\rightarrow$ random number between 0 and 1
S_{max} $\rightarrow$ maximum step size allowed to move a frog after being infected.

The new position for the worst frog can now be evaluated as in Eq 7.4:

$$U(q) = P_W + S \tag{7.4}$$

If the new position is within the search space, then we would move ahead by evaluating the objective function value for this new position given by 'f(q)'. If the new 'f(q)' is better than the previous value, i.e., the objective function value of the worst frog, than replace the worst frog with the new one in the submemeplex. Now replace the frogs in the submemeplex 'Z' back in their original positions of the memeplex and continue with memetic evolution for the concerned memeplex. In case, the new frog position does not lie in the search space or the new objective function value is not better than the previous value, then utilise Eq 7.5 to upgrade the position of the worst frog in the submemeplex.

$$\begin{aligned} S &= \min\{\text{int}[\text{rand}(0,1)(P_X - P_W)], S_{\max}\} \quad \text{for a positive step} \\ &= \max\{\text{int}[\text{rand}(0,1)(P_X - P_W)], -S_{\max}\} \quad \text{for a negative step} \end{aligned} \tag{7.5}$$

Eq 7.5 is similar to Eq 7.3, except that 'P_B' is replaced by 'P_X'. 'P_X' represents the best frog in the entire population. The new frog position shall be again found through Eq 7.4. If the new frog position lies within the search space, its objective function value is found. In case the new objective function value is better than the previous one then replace the worst frog with the new position. Next replace the frogs in the submemeplex 'Z' back in their original positions of the memeplex and continue with memetic evolution for the concerned memeplex. In case, the new frog position is not within the search space or the new objective function value is not better than the previous one, then, a frog is randomly generated within the search space to replace the worst frog. After replacement of the worst frog, submemeplex frogs shall be replaced in their original positions in the memeplex and the memetic evolution continues.

After passing a particular memeplex through multiple memetic evolution, the same process is repeated for the remaining memeplexes. Once all the memeplexes go through the memetic evolution, all the frogs in different memeplexes are again gathered in 'X'. This process is done to include global exploration in SFLA. The matrix 'X' is again sorted based on the objective function values of the frogs. Again memeplexes are created and memetic evolution is performed on each memeplex. This process is repeated until convergence is achieved. In SFLA [41], convergence is achieved when the frog carrying the best memetic pattern does not change for a prespecified number of iterations. The process of SFLA is quite complex and is explained through algorithm and flowchart in the following sections.

7.4 ALGORITHM AND PSEUDOCODE

The algorithms presented in this section are based on the flowcharts of the next section. In other words, the algorithms / flowcharts of SFLA are divided into smaller processes for better understanding. Also some of the minor steps may be skipped as the SFLA algorithm is quite complex in nature. The Algorithm 7.1 gives the overview of the whole process involved in SFLA. The algorithm starts with input of required data and random generation of frogs within the swamp area / limits. In the next step, the objective function values of each frog is evaluated and the frogs and their repsective objective function values are collaborated to form the 'X' matrix.

Algorithm 7.1 Shuffled Frog Leaping Algorithm

```
START
Input Required Data
for 1 to Population Size do
   Generate Frogs
   if Meme > Max Limit or Meme < Min Limit then
      Meme = Limit
   end if
end for
for 1 to Population Size do
   Obtain Objective Function Value
end for
Create X = [U(i) f(i)]
repeat
   Sort X
   Create Memeplexes
   Perform Memetic Evolution
   Combine Memeplexes
until Check Convergence
Display Result
STOP
```

The main loop of SFLA starts after the formation of matrix 'X'. This loop is repeated until the convergence is achieved. In one loop / iteration, the 'X' matrix is first sorted, then it is divided into various memeplexes, after which memetic evolution is performed and then the memeplexes are recombined. After the end of one iteration, the convergence may be checked. The steps presented in the iteration are further expanded in the following algorithms.

Algorithm 7.2 gives the algorithm to form memeplexes from 'X'. The frogs, which are arranged in a sorted manner in 'X', are distributed into memeplexes as explained in Fig 7.1. A nested loop is utilized to achieve the distribution of frogs into memplexes.

Algorithm 7.2 Creating Memeplexes

for 1 **to** m **do**
  $Y^i = [\ \ ]$
  for 1 **to** n **do**
    $Y^i = [Y^i; U(i+m(j-1), f((i+m(j-1)]$
  end for
end for

In Algorithm 7.2, it is assumed that the outer loop uses variable 'i' and inner loop uses variable 'j'. The outer loop represents the formation of one memeplex and is therefore repeated 'm' times. The inner loop populates 'n' frogs in each memeplex using Eq 7.1.

After the creation of memeplexes, the memetic evolution is started for each memeplex. In a memetic evolution, the first step is to select certain frogs from the memeplex and improve the position of the worst frog. Algorithm 7.3 gives the process to form submemeplexes based on triangular probability distribution method.

Algorithm 7.3 Creating SubMemeplexes

for 1 **to** q **do**
  Generate random number
  Select Frog from Memeplex based on Random Number
  if Frog $\in$ Z **then**
    Repeat Loop
  else
    Add Frog to Z
  end if
end for
Sort Z
Find P_B and P_W

To create submemeplex 'Z', a loop is executed 'q' times, to generate random numbers and to select frogs based on these random numbers through triangular probability distribution. In the loop, we would like to check if the frog is already a part of 'Z' or not. If it is, we repeat the loop or else we add the frog to 'Z'. Once the 'Z' matrix is formed, next we sort the matrix and assign the frog with best objective function value as 'P_B' and that with worst objective function value as 'P_W'.

The position of the worst frog now needs to be improved, which is achieved through Algorithm 7.4.

To improve the position of the worst frog, the step size is first evaluated using 'P_B'. With this step size, the new position of the frog 'U_q' can be obtained. It is evaluated whether the frog's new position lies in the search space 'Ω' or not. If it lies in the search space then the objective function value is found. The objective function value is compared with the previous objective function value, in case a better solution / position is obtained, the old position is replaced with the new one.

If the new position is not within the search space or the new objective function value does not provide a better solution, a new step size is evaluated with the help of 'P_X'. Again new 'U_q' is found and the same process is repeated, i.e., check whether the new position is in the search space and the new objective function value is better than the old one. If the new position is better than the old one and it lies in the search space then the old position of the worst frog is replaced with the new one. Otherwise, a random frog is generated within the search space and it replaces the old position of the worst frog.

The memetic evolution process shown in Algorithms 7.3 and 7.4 are repeated for a number of times for a single memeplex.

As said earlier, some of the steps are not displayed in the algorithms; however, it should be taken care that the 'P_X' needs to be updated for every change in a frog.

Algorithm 7.4 Improvement in Worst Frog Position

Find Step Size S using P_B (Eq 7.3)
Find U_q (Eq 7.4)
if $U_q \in \Omega$ **then**
  Find f(q)
  if $f(q)_{new} < f(q)_{old}$ **then**
    Replace $U_{q,old}$ with $U_{q,new}$
  else
    GoTo A
  end if
else
  Label A
  Find Step Size S using P_X (Eq 7.5)
  Find U_q (Eq 7.4)
  if $U_q \in \Omega$ **then**
    Find f(q)
    if $f(q)_{new} < f(q)_{old}$ **then**
      Replace $U_{q,old}$ with $U_{q,new}$
    else
      GoTo B
    end if
  else
    Label B
    Generate Random Frog
    Replace $U_{q,old}$ with Random Frog
  end if
end if

7.5 FLOWCHART

The SFLA method is a very complex EOA. It is not possible to conveniently explain the method through a single flowchart, thus the process has been divided into different processes to get a better idea of SFLA. The flowchart in Fig 7.3, gives the overall process of SFLA. The major steps mentioned in the flowchart are: generation of random frogs, then sorting of frogs into X matrix, followed by creation of memeplexes and memetic evolution. After the completion of memetic evolution, the memeplexes are recombined, and the convergence is checked. In case the process has converged and the optimal solution has been found, the SFLA search process is stopped. However, in case, the convergence has not been reached, then the frogs are again sorted and the whole iteration is repeated.

It is necessary here for the reader to understand that these are broad steps in SFLA. In each step there might be multiple steps, loops, decision making statements and so on. Let us try and explore some of these steps in detail. A generation of random frogs is similar to generation of random particles, etc. Also sorting of frogs based on their objective function value can be done through a single command in many of the programming tools available, thus the techniques used for sorting of frogs is out of the scope of this book.

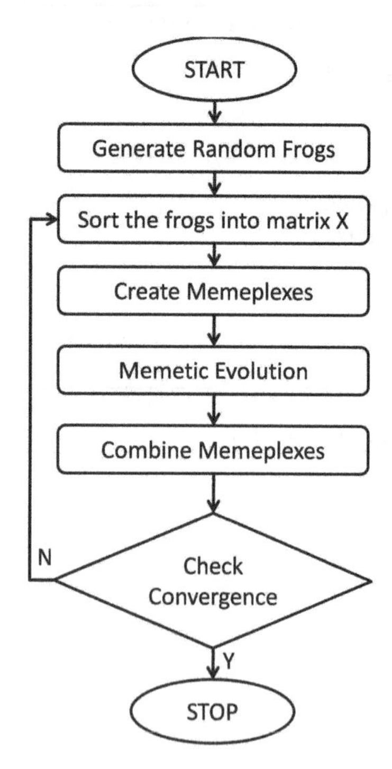

Figure 7.3: SFLA Flowchart

Let us expand the creation of memeplexes and see how it can be achieved. The flowchart for creating the memeplexes is shown in Fig 7.4. For generating memeplexes from the sorted matrix 'X', we will be requiring a nested loop as shown in the figure with the outer loop repeating up to 'm', i.e., the number of memeplexes and the inner loop repeating 'n' times, i.e., number of frogs in a memeplex. Before starting to fill a memeplex 'Y^i' with frogs, it would be required to clear the memeplex of any previous content as shown in the flowchart. The frogs are distributed into memeplexes through the formula displayed in the flowchart.

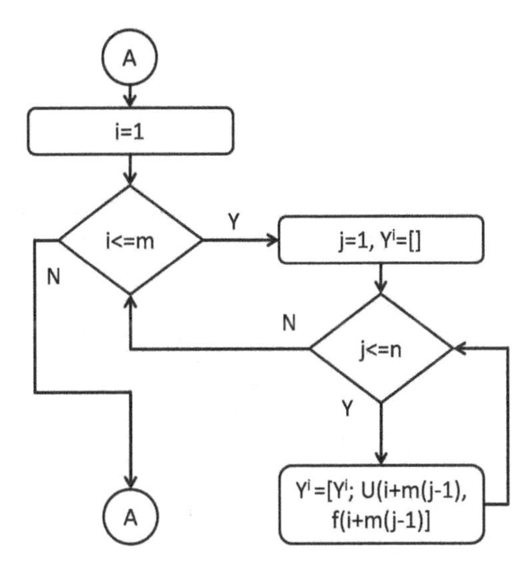

Figure 7.4: Creating Memeplexes

After creation of memeplexes, the next step is to start memetic evolution for each memeplex. Thus, even though in Fig 7.3, memetic evolution is a single block, in coding it needs to be replaced with a loop for each memeplex. In other words, memetic evolution should take place for each memeplex for a certain number of times. Memetic evolution itself is a big process and is divided into two flowcharts for easier understanding. The two parts of memetic evolution are : creation of submemeplexes and movement of worst frogs in a submemeplex to a better position. The creation of submemeplex is given in Fig 7.5.

To create submemeplex 'Z', we create a loop for 'q' times where 'q' is the number of frogs to be included in a submemeplex. A submemeplex is created through triangular probability distribution method, which allows a frog with a better objective function value to be given a better chance to be a part of the submemeplex. A random number between 0 and 1 is first generated, then based on the random number a frog from the respective memeplex is selected, as explained earlier through Fig 7.2. It is then required to be assessed if the frog is already a part of the submemeplex or not. In case the selected frog is already present in the submemeplex, then the loop needs to be repeated for getting a different frog from the memeplex, otherwise the selected frog is added to the submemeplex and the loop continues. It should be noted here, that nothing has been specifically mentioned about duplicate entry of frogs in submemeplex in the references, it is however natural to take this into consideration, since duplicate entries will defeat the purpose of creating the submemeplex. Still it is up to the reader / user to allow duplicate entries of frogs into a submemeplex.

Once submemeplex has been formed, the submemeplex 'Z' is sorted. The best frog 'P_B' and the worst frog 'P_W' within the submemeplex are found. This ends the first part of memetic evolution.

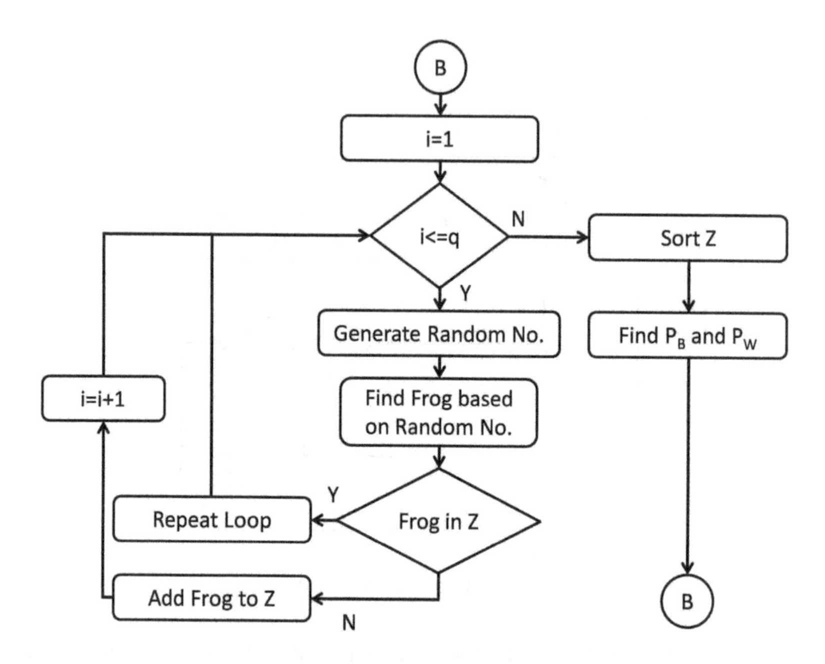

Figure 7.5: Creation of Submemeplex

In the second part of memetic evolution, the worst frog 'P_W' in the submemeplex gets infected from another frog 'P_B' and moves to a better position as shown by the flowchart in Fig 7.6. To move the worst frog to a better position, the step size 'S' is first evaluated using 'P_B' in Eq 7.3. Further the new position 'U_q' is found using Eq 7.4. If the new position of the worst frog is within the search space then its objective function value 'f_q' is found. If 'f_q' is better than the objective function value of the worst frog in the submemeplex, then the worst frog is replaced by the new position. After replacing the worst frog with a better solution, the frogs of the submemeplex are placed back into their respective memeplex.

On the other hand, in case the new position of the frog lies outside the search space 'Ω' or 'f_q' is not an improvement of the objective function value of the worst frog, then the step size is again evaluated using 'P_X' in Eq 7.5. A similar process is repeated and the new position is found, then it is checked whether it is within the search space or not. If the new position is still not within the search space then a random frog is generated and is used to replace the worst frog. If the new position is within the search space then the objective function value is found. If the new objective function value shows some improvement over the existing objective function value then the new frog replaces the worst frog, otherwise a random frog is generated and it replaces the worst frog. Thus the worst frog shall be replaced with either a better solution or a random one. As a last step, the frogs of the submemeplex are placed back in their respective memeplex.

It can be seen that there is a lot of repetitiveness in the explanation, as multiple conditions are implemented and their results end up at the same positions. Anyhow, it can be observed that the new position of the worst frog can be obtained in three ways: (i) 'P_B', (ii) 'P_X' or (iii) random frog. Once the new position of the worst frog has been decided, the frogs in the submemeplex are replaced into their respective positions in the parent memeplex.

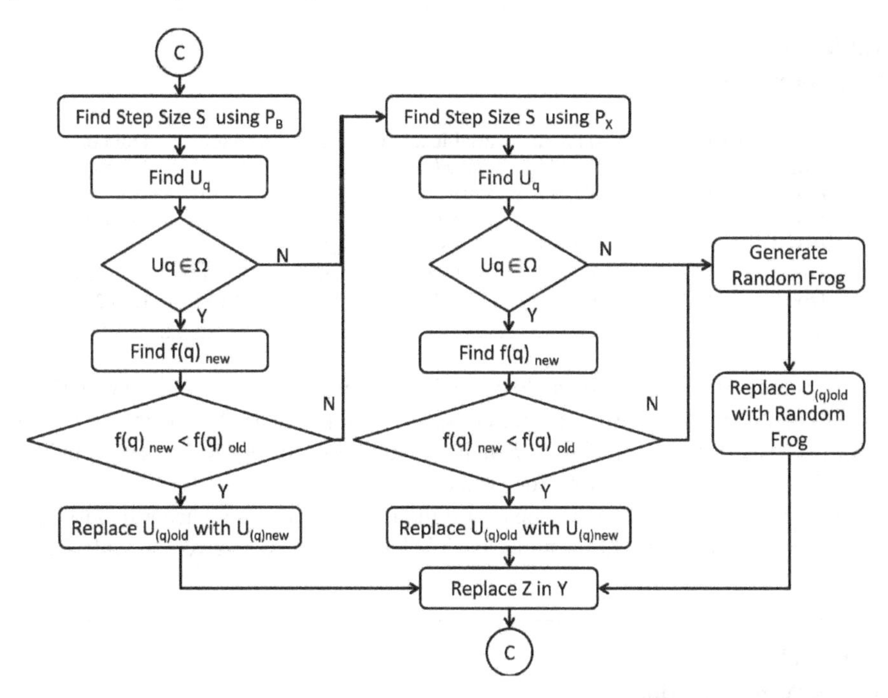

Figure 7.6: Improvement in Worst Frog Position

The memetic evolution represented by flowcharts in Figs 7.5 and 7.6, are repeated for a specified number of iterations for each memeplex. After the memetic evolution / local search is over, all the memeplexes are recombined in the matrix 'X'. The search process should now confirm the occurrence of convergence, in case the convergence has occurred, the search process should stop. However, if the convergence is not met, the 'X' matrix is again sorted and the same process is repeated.

7.6 EXAMPLE

The example section is used in this book to solve an optimization technique step-by-step for two iterations. In SFLA, there are a plenty of variations which are possible, however, all of them may not be reflected in the operation of the first two iterations, like generation of a random frog. Still let's see how the calculations are performed along with their steps in SFLA. The Griewank problem is solved in this section, whose details can be obtained from Eqs 2.5 and 2.9.

In SFLA, frogs are possible soutions for the given problem. Similar to other chapters, we assume the following set of frogs to be randomly generated and as a starting point for the search process. Table 7.1 gives the frogs along with their objective function value, since objective function value are required for sorting.

Table 7.1
Frogs in the Swamp

	Variable 1	Variable 2	Variable 3	Variable 4	Variable 5	Obj Fun
Frog 1	141.8811	−104.656	−397.606	211.7688	561.0059	138.1064
Frog 2	−16.0475	−277.517	74.94419	247.9742	391.221	75.3538
Frog 3	503.4618	202.9188	51.23069	356.5671	321.7123	132.9249
Frog 4	−80.9499	−329.007	−180.324	77.46379	397.8965	78.7628
Frog 5	471.3352	−152.64	86.91336	386.1857	−460.428	154.5424
Frog 6	−556.071	286.2077	118.7875	249.8144	106.6543	120.6237

We start with 6 frogs, and would be having two memeplex (i.e., m = 2) with three frogs (i.e., n = 3) in each memeplex. Let us also assume that a sub memeplex 'Z' will have two frogs (i.e., q = 2). Even though the numbers are very less, they should be enough to show the calculations for SFLA.

The first step in SFLA would be to sort the frogs in the swamp according to their objective function values to form matrix 'X'. The matrix 'X' is given in Table 7.2.

Table 7.2
Matrix 'X' for Iteration 1

	Variable 1	Variable 2	Variable 3	Variable 4	Variable 5	Obj Fun
Frog 1	−16.0475	−277.517	74.94419	247.9742	391.221	75.35377
Frog 2	−80.9499	−329.007	−180.324	77.46379	397.8965	78.76276
Frog 3	−556.071	286.2077	118.7875	249.8144	106.6543	120.6237
Frog 4	503.4618	202.9188	51.23069	356.5671	321.7123	132.9249
Frog 5	141.8811	−104.656	−397.606	211.7688	561.0059	138.1064
Frog 6	471.3352	−152.64	86.91336	386.1857	−460.428	154.5424

The frog numbering was not important when they were randomly generated, thus we retain the frog numbers as it is in the first column of the table. The objective function value can be observed to be in ascending order.

The frogs are now to be divided into memeplexes as shown in Fig 7.1 and Eq 7.1. Since we are going to create two memeplexes with three frogs each, it would be an easy task. Still let's see how the distribution happens with the Eq 7.1. The frogs in memeplex 1 and 2 would be calculated as:

First member of memeplex 1:

$$Y^1 = [U(1+2(1-1), f(1+2(1-1)] = [U(1), f(1)]$$

where i =1 (memeplex number), m = 2 (number of memeplexes) and j =1 (first member)

Similarly, second member would be found as:

$$Y^1 = [U(1+2(2-1), f(1+2(2-1)] = [U(3), f(3)]$$

where j =2

And the last member would be:

$$Y^1 = [U(1+2(3-1), f(1+2(3-1)] = [U(5), f(5)]$$

For the second memeplex, the calculation would be given as:

$$Y^2 = [U(2+2(1-1), f(2+2(1-1)] = [U(2), f(2)]$$
$$Y^2 = [U(2+2(2-1), f(2+2(2-1)] = [U(4), f(4)]$$
$$Y^2 = [U(2+2(3-1), f(2+2(3-1)] = [U(6), f(6)]$$

The distribution of frogs is completed and now the memeplexes would be represented as given in Table 7.3.

Table 7.3
Memeplexes Iteration 1

	Variable 1	Variable 2	Variable 3	Variable 4	Variable 5	Obj Fun
Y^1						
Frog 1	−16.0475	−277.517	74.94419	247.9742	391.221	75.35377
Frog 3	−556.071	286.2077	118.7875	249.8144	106.6543	120.6237
Frog 5	141.8811	−104.656	−397.606	211.7688	561.0059	138.1064
Y^2						
Frog 2	−80.9499	−329.007	−180.324	77.46379	397.8965	78.76276
Frog 4	503.4618	202.9188	51.23069	356.5671	321.7123	132.9249
Frog 6	471.3352	−152.64	86.91336	386.1857	−460.428	154.5424

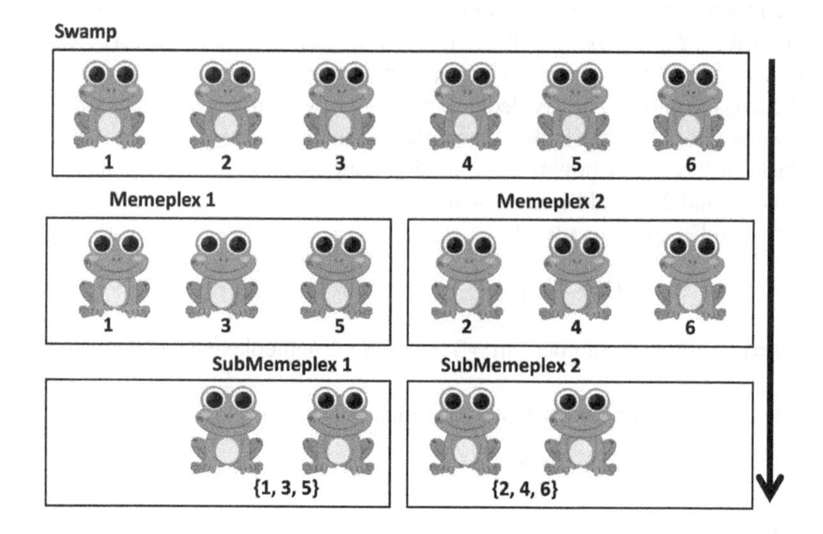

Figure 7.7: Swamp to Memeplexes to SubMemeplexes

The distribution of the frogs from a swamp to memeplexes and then to submemeplexes in this example is shown in Fig 7.7.

'P_X' would be the best frog in the entire swamp and has to be recorded all the time.

One memeplex has 3 frogs out of which we need to select 2 frogs. For this triangular probability distribution is used, the formula for which is given in Eq 7.2. The probabilities assigned to different frogs in the memeplex is evaluated as:

$$
\begin{aligned}
p_1 &= \frac{2(3+1-1)}{3(3+1)} = 0.5 \\
p_2 &= \frac{2(3+1-2)}{3(3+1)} = 0.3333 \\
p_3 &= \frac{2(3+1-3)}{3(3+1)} = 0.1667
\end{aligned}
$$

In the above equation as compared to Eq 7.2, n shall be taken as 3 and j will vary from 1 to 3 for different frogs in the memeplex. It is but obvious that the frogs in the memeplex shall be arranged in sorted order. The triangular probability distribution amongst the frogs in the memeplex can be shown through Fig 7.8.

In the next step we form submemeplex, in which random numbers will be generated to select frogs from a memeplex. The frog in whose area the random number lies, that frog will be selected for the submemeplex. For frog 1, the range is from 0

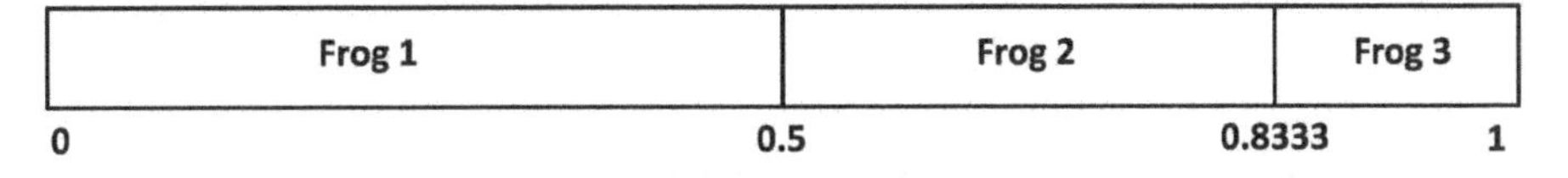

Figure 7.8: Triangular Probability Distribution

to 0.5, for frog 2 it is 0.5–0.8333 and for frog 3 it is 0.8333–1, irrespective of the memeplex.

The random numbers generated for creating submemeplexes are: 0.8147 and 0.9058. In case, the random numbers point to the same frog in the memeplex, then another random number will have to be generated to select a different frog. According to Fig 7.8 and triangular probability distribution, the frogs selected for submemeplex are – 2nd and 3rd, respectively from memeplex 1. Since these frogs are selected in proper order, sorting may not be required, otherwise if the 3rd frog is selected first and then the 2nd, sorting will be required.

The submemeplex matrix 'Z' can be given as in Table 7.4:

Table 7.4
SubMemeplex from Memeplex 1 in Iteration 1

	Variable 1	Variable 2	Variable 3	Variable 4	Variable 5	Obj Fun
Frog 3	−556.071	286.2077	118.7875	249.8144	106.6543	120.6237
Frog 5	141.8811	−104.656	−397.606	211.7688	561.0059	138.1064

In the next step, the matrix 'Z' is sorted, however, it is not required here as it is already sorted. Also from the matrix we can observe that 'P_B' is Frog 3 and 'P_W' is Frog 5 for the current submemeplex.

Now to improve the position of the worst frog through Eq 7.3, we need to generate random numbers. These randomly generated numbers are:

0.127, 0.9134, 0.6324, 0.0975, 0.2785

It is assumed that 'S_{Max}' is 50 for positive step and −50 from negative step. Let us calculate the step size for each variable in the worst frog 'P_W'.

Note: 1) In Eqs 7.3 and 7.5, the calculation of step size involves "int", which means the step size should be an integer and decimals will not be allowed. But then this will call for a minimum step size of 1. Since our variable are continuous in nature, the "int" in these equations are ignored.

2) There is no mention as to whether the step size should be calculated in a single step or each variable should be evaluated separately. But considering Eqs 7.3 and 7.5, it is observed that the step size can be positive or negative. The whole step vector cannot be classified as positive or negative, thus it is concluded that the step size should be calculated for each individual variable separately.

Variable 1:

$$P_B(1) - P_W(1) = -556.071 - 141.8811 = -697.9521$$
$$rand() * (P_B(1) - P_W(1)) = 0.127 * (-697.9521) = -88.6399$$
$$S(1) = max(-88.6399, -50) = -50$$

The random values are given above whereas the remaining values are taken from Table 7.4.

Variable 2:

$$P_B(2) - P_W(2) = 286.2077 - (-104.656) = 390.8637$$
$$rand() * (P_B(2) - P_W(2)) = 0.9134 * 390.8637 = 357.0149$$
$$S(2) = min(357.0149, 50) = 50$$

Similarly for other variables, the calculations will be as:

Variable 3:

$$P_B(3) - P_W(3) = 118.7875 - (-397.606) = 516.3935$$
$$rand() * (P_B(3) - P_W(3)) = 0.6324 * 516.3935 = 326.5672$$
$$S(3) = min(326.5672, 50) = 50$$

Variable 4:

$$P_B(4) - P_W(4) = 249.8144 - 211.7688 = 38.0456$$
$$rand() * (P_B(4) - P_W(4)) = 0.0975 * 38.0456 = 3.7094$$
$$S(4) = min(3.7094, 50) = 3.7094$$

Variable 5:

$$P_B(5) - P_W(5) = 106.6543 - 561.0059 = -454.3516$$
$$rand() * (P_B(5) - P_W(5)) = 0.2785 * (-454.3516) = -126.5369$$
$$S(5) = max(-126.5369, -50) = -50$$

The step size is summarized in Table 7.5.

Table 7.5
Step Size for Memeplex 1 in Iteration 1

	Variable 1	Variable 2	Variable 3	Variable 4	Variable 5
S	−50	50	50	3.7094	−50

Adding this step to the worst frog in Table 7.4, we get the new position for the worst frog '$U_{q,new}$', as given in Table 7.6.

Table 7.6
New Position of Worst Frog for Memeplex 1 in Iteration 1

	Variable 1	Variable 2	Variable 3	Variable 4	Variable 5
Frog 5	91.88112	−54.6557	−347.606	215.4798	511.0059

The objective function value for this new position of the worst frog is found to be: 110.7932. Since this objective function value is better than the previous one, the worst frog is replaced with the new position. The submemeplex 'Z' can now be represented as in Table 7.7:

Table 7.7
SubMemeplex from Memeplex 1 in Iteration 1 after Memetic Evolution

	Variable 1	Variable 2	Variable 3	Variable 4	Variable 5	Obj Fun
Frog 3	−556.071	286.2077	118.7875	249.8144	106.6543	120.6237
Frog 5	91.88112	−54.6557	−347.606	215.4798	511.0059	110.7932

Once the new frog position has been evaluated, the submemeplex should be placed back in its respective memeplex. In this case, the memeplex is 'Y^1', represented in Table 7.8.

Table 7.8
Memeplex Y^1 in Iteration 1

	Variable 1	Variable 2	Variable 3	Variable 4	Variable 5	Obj Fun
Frog 1	−16.0475	−277.517	74.94419	247.9742	391.221	75.35377
Frog 3	−556.071	286.2077	118.7875	249.8144	106.6543	120.6237
Frog 5	91.88112	−54.6557	−347.606	215.4798	511.0059	110.7932

The memplex should be sorted immediately, so that the triangular probability distribution is properly implemented, as given in Table 7.9. After sorting, Frog 5 moves to the second position in mememplex 1.

The process of creating submemeplex, finding step size, improving frog position and replacing submemeplex back into its respective memeplex should be repeated a number of times. Since this section is just an example, the process is performed only once and we move on to the second memeplex.

Two random numbers will be first generated to select two frogs from memeplex 2. The random numbers generated are: 0.5469 and 0.9575. Thus, the 2nd and 3rd frog

Table 7.9
Memeplex Y^1 after Sorting in Iteration 1

	Variable 1	Variable 2	Variable 3	Variable 4	Variable 5	Obj Fun
Frog 1	−16.0475	−277.517	74.94419	247.9742	391.221	75.35377
Frog 5	91.88112	−54.6557	−347.606	215.4798	511.0059	110.7932
Frog 3	−556.071	286.2077	118.7875	249.8144	106.6543	120.6237

will be selected into the submemeplex 'Z' as the random numbers lie in these frogs area (Refer Fig 7.8). The submemeplex is shown in Table 7.10. Since the random numbers are so generated that the SubMemeplex is already sorted, we need not perform sorting on it. From the table it can be observed that Frog 6 is the worst frog 'P_W' and Frog 4 is the best frog 'P_B', in the SubMemeplex.

Table 7.10
SubMemeplex from Memeplex 2 in Iteration 1

	Variable 1	Variable 2	Variable 3	Variable 4	Variable 5	Obj Fun
Frog 4	503.4618	202.9188	51.23069	356.5671	321.7123	132.9249
Frog 6	471.3352	−152.64	86.91336	386.1857	−460.428	154.5424

Now to generate random values to calculate the step of the worst frog in SubMemeplex. The random numbers generated are: 0.9649, 0.1576, 0.9706, 0.9572 and 0.4854. One random number is generated for each variable. The calculations for each variable is presented below:

(Values are taken from Table 7.10).

Variable 1:

$$P_B(1) - P_W(1) = 503.4618 - 471.3352 = 32.1266$$
$$rand() * (P_B(1) - P_W(1)) = 0.9649 * 32.1266 = 30.9989$$
$$S(1) = min(30.9989, 50) = 30.9989$$

Variable 2:

$$P_B(2) - P_W(2) = 202.9188 - (-152.64) = 355.5588$$
$$rand() * (P_B(2) - P_W(2)) = 0.1576 * 355.5588 = 56.0361$$
$$S(2) = min(56.0361, 50) = 50$$

Variable 3:

$$P_B(3) - P_W(3) = 51.2307 - 86.9134 = -35.6827$$
$$rand() * (P_B(3) - P_W(3)) = 0.9706 * -35.6827 = -34.6336$$
$$S(3) = max(-34.6336, -50) = -34.6336$$

Variable 4:

$$P_B(4) - P_W(4) = 356.5671 - 386.1857 = -29.6186$$
$$rand() * (P_B(4) - P_W(4)) = 0.9572 * -29.6186 = -28.3509$$
$$S(4) = max(-28.3509, -50) = -28.3509$$

Variable 5:

$$P_B(5) - P_W(5) = 321.7123 - (-460.428) = 782.1403$$
$$rand() * (P_B(5) - P_W(5)) = 0.4854 * 782.1403 = 379.6509$$
$$S(5) = min(379.6509, 50) = 50$$

Note: The values in the tables are obtained through MATLAB which uses more number of decimals and hence there might be some minor differences in the values listed in the tables.

The step size is summarized in Table 7.11.

Table 7.11
Step Size for Memeplex 2 in Iteration 1

	Variable 1	Variable 2	Variable 3	Variable 4	Variable 5
S	30.9989	50	−34.6336	−28.3509	50

Adding this step to the worst frog in Table 7.10, we get the new position for the worst frog '$U_{q,new}$', as given in Table 7.12.

Table 7.12
New Position of Worst Frog for Memeplex 2 in Iteration 1

	Variable 1	Variable 2	Variable 3	Variable 4	Variable 5
Frog 6	502.3338	−102.64	52.28002	357.8357	−410.428

The objective function value of the new position of the worst frog is found to be 141.4575. The SubMemeplex can now be represented as given in Table 7.13

Table 7.13
SubMemeplex from Memeplex 2 in Iteration 1 after Memetic Evolution

	Variable 1	Variable 2	Variable 3	Variable 4	Variable 5	Obj Fun
Frog 4	503.4618	202.9188	51.23069	356.5671	321.7123	132.9249
Frog 6	502.3338	−102.64	52.28002	357.8357	−410.428	141.4575

Table 7.14
Memeplex Y^2 in Iteration 1

	Variable 1	Variable 2	Variable 3	Variable 4	Variable 5	Obj Fun
Frog 2	−80.9499	−329.007	−180.324	77.46379	397.8965	78.76276
Frog 4	503.4618	202.9188	51.23069	356.5671	321.7123	132.9249
Frog 6	502.3338	−102.64	52.28002	357.8357	−410.428	141.4575

The elements of the SubMemeplex has to be replaced in the respective Memeplex, i.e., 'Y^2'. The memeplex can now be represented as in Table 7.14.

The frogs are already arranged in the sorted order according to their objective function values in memeplex 2 and thus another table is not used to show the sorted memeplex. As mentioned earlier, this process should be repeated a number of times for a single memeplex, but for better understanding of the process the memetic evolution of memeplex 2 is performed only for a single iteration. After the memetic evolution has been performed, all the memeplex are compiled into a single matrix, this matrix is shown in Table 7.15. It can be observed that the position of the best frog in the whole swamp is yet to change its position, and the frog 'P_X' remains the same without getting replaced.

Table 7.15
Frogs in the Swamp after Iteration 1

	Variable 1	Variable 2	Variable 3	Variable 4	Variable 5	Obj Fun
Y^1						
Frog 1	−16.0475	−277.517	74.94419	247.9742	391.221	75.35377
Frog 5	91.88112	−54.6557	−347.606	215.4798	511.0059	110.7932
Frog 3	−556.071	286.2077	118.7875	249.8144	106.6543	120.6237
Y^2						
Frog 2	−80.9499	−329.007	−180.324	77.46379	397.8965	78.76276
Frog 4	503.4618	202.9188	51.23069	356.5671	321.7123	132.9249
Frog 6	502.3338	−102.64	52.28002	357.8357	−410.428	141.4575

Now the frogs are sorted to form the matrix 'X', as shown in Table 7.16:

Table 7.16
Matrix 'X' for Iteration 2

	Variable 1	Variable 2	Variable 3	Variable 4	Variable 5	Obj Fun
Frog 1	−16.0475	−277.517	74.94419	247.9742	391.221	75.35377
Frog 2	−80.9499	−329.007	−180.324	77.46379	397.8965	78.76276
Frog 5	91.88112	−54.6557	−347.606	215.4798	511.0059	110.7932
Frog 3	−556.071	286.2077	118.7875	249.8144	106.6543	120.6237
Frog 4	503.4618	202.9188	51.23069	356.5671	321.7123	132.9249
Frog 6	502.3338	−102.64	52.28002	357.8357	−−410.428	141.4575

As shown earlier, the matrix 'X' shall be bifurcated into two mememplexes, which is presented in Table 7.17.

Table 7.17
Formation of Memeplexes for Iteration 2

Memeplex 1 (Y^1)	Memeplex 2 (Y^2)
Frog 1	Frog 2
Frog 5	Frog 3
Frog 4	Frog 6

The frogs 4 and 3 interchange their positions in the memeplexes. In the next step the submemeplex should be created from the first memeplex. The random number generated to select these frogs are: 0.8003 and 0.1419. The frogs selected based on these random numbers would be 5 and 1, respectively. The submemeplex so formed is presented in Table 7.18.

Table 7.18
SubMemeplex from Memeplex 1 in Iteration 2

	Variable 1	Variable 2	Variable 3	Variable 4	Variable 5	Obj Fun
Frog 5	91.88112	−54.6557	−347.606	215.4798	511.0059	110.7932
Frog 1	−16.0475	−277.517	74.94419	247.9742	391.221	75.35377

In this case, we find that the submemeplex is not sorted and needs to be sorted according to the objective function values. This is shown in Table 7.19. Sorting of frogs in a memeplex is important since: (1) 'P_B' and 'P_W' are to be ascertained and

(2) Submemeplex will not always have two frogs. For more number of frogs, sorting will become necessary to ascertain 'P_B' and 'P_W'.

Table 7.19
Sorted SubMemeplex from Memeplex 1 in Iteration 2

	Variable 1	Variable 2	Variable 3	Variable 4	Variable 5	Obj Fun
Frog 1	−16.0475	−277.517	74.94419	247.9742	391.221	75.35377
Frog 5	91.88112	−54.6557	−347.606	215.4798	511.0059	110.7932

Frog 1 is 'P_B' and Frog 5 is 'P_W', for the above SubMemeplex. The step size is found with the help of random numbers generated for each input variable, which are: 0.4218, 0.9157, 0.7922, 0.9595 and 0.6557. The calculations to find the step size are shown before and are not repeated here. The step size is shown in Table 7.20.

Table 7.20
Step Size for Memeplex 1 in Iteration 2

	Variable 1	Variable 2	Variable 3	Variable 4	Variable 5
S	−45.5201	−50	50	31.17815	−50

The new position of the worst frog in Memeplex 1 for Iteration 2 is now presented in Table 7.21.

Table 7.21
New Position of Worst Frog for Memeplex 1 in Iteration 2

	Variable 1	Variable 2	Variable 3	Variable 4	Variable 5
Frog 5	46.36102	−104.656	−297.606	246.658	461.0059

The objective function value of the new position of the worst frog is found to be 94.7786. Table 7.22 gives the new Submemeplex for Memeplex 1 in Iteration 2. Sorting will not be required as the frogs are already sorted in order.

The frogs in the submemeplex are replaced in their original position of their respective memeplex 1. The memeplex 1 is represented in Table 7.23. The memeplex remains sorted after replacement and therefore sorting is not required.

Now to move towards the last iteration for last memeplex of this example. The random numbers generated for selection of frogs in the submemeplex are: 0.0357 and 0.8491. Based on these random numbers, frogs 2 and 6 are selected for the

Table 7.22
SubMemeplex from Memeplex 1 in Iteration 2 after Memetic Evolution

	Variable 1	Variable 2	Variable 3	Variable 4	Variable 5	Obj Fun
Frog 1	−16.0475	−277.517	74.94419	247.9742	391.221	75.35377
Frog 5	46.36102	−104.656	−297.606	246.658	461.0059	94.77863

Table 7.23
Memeplex Y^1 in Iteration 2

	Variable 1	Variable 2	Variable 3	Variable 4	Variable 5	Obj Fun
Frog 1	−16.0475	−277.517	74.94419	247.9742	391.221	75.35377
Frog 5	46.36102	−104.656	−297.606	246.658	461.0059	94.77863
Frog 4	503.4618	202.9188	51.23069	356.5671	321.7123	132.9249

submemeplex. The new submemeplex is represented in Table 7.24. The frogs in the submemeplex are sorted and it does not require sorting. Frog 2 is 'P_B' and Frog 6 is 'P_W' for the given submemeplex.

Table 7.24
SubMemeplex from Memeplex 2 in Iteration 2

	Variable 1	Variable 2	Variable 3	Variable 4	Variable 5	Obj Fun
Frog 2	−80.9499	−329.007	−180.324	77.46379	397.8965	78.76276
Frog 6	502.3338	−102.64	52.28002	357.8357	−410.428	141.4575

A set of random numbers are generated to find the step size for the movement of the worst frog. These random numbers are: 0.934, 0.6787, 0.7577, 0.7431 and 0.3922.

The step size for the worst frog based on the values available are found as shown in Table 7.25.

The new position of the worst frog is obtained as shown in Table 7.26

The objective function value for this new position of the worst frog in submemeplex is found to be 114.0829. Since there is an improvement in the frog position, the frog 6 is replaced by the new position.

The submemeplex is not represented in Table 7.27.

The frogs in the submemeplex needs to be now replaced in the memeplex. The memeplex after replacement of frogs is shown in Table 7.28.

The memeplex can be sorted and shown in Table 7.29.

Table 7.25
Step Size for Memeplex 2 in Iteration 2

	Variable 1	Variable 2	Variable 3	Variable 4	Variable 5
S	−50	−50	−50	−50	50

Table 7.26
New Position of Worst Frog for Memeplex 2 in Iteration 2

	Variable 1	Variable 2	Variable 3	Variable 4	Variable 5
Frog 6	452.3338	−152.64	2.280021	307.8357	−360.428

Table 7.27
SubMemeplex from Memeplex 2 in Iteration 2 after Memetic Evolution

	Variable 1	Variable 2	Variable 3	Variable 4	Variable 5	Obj Fun
Frog 2	−80.9499	−329.007	−180.324	77.46379	397.8965	78.76276
Frog 6	452.3338	−152.64	2.280021	307.8357	−360.428	114.0829

Table 7.28
Memeplex Y^2 in Iteration 2

	Variable 1	Variable 2	Variable 3	Variable 4	Variable 5	Obj Fun
Frog 2	−80.9499	−329.007	−180.324	77.46379	397.8965	78.76276
Frog 3	−556.071	286.2077	118.7875	249.8144	106.6543	120.6237
Frog 6	452.3338	−152.64	2.280021	307.8357	−360.428	114.0829

Table 7.29
Sorted Memeplex Y^2 in Iteration 2

	Variable 1	Variable 2	Variable 3	Variable 4	Variable 5	Obj Fun
Frog 2	−80.9499	−329.007	−180.324	77.46379	397.8965	78.76276
Frog 6	452.3338	−152.64	2.280021	307.8357	−360.428	114.0829
Frog 3	−556.071	286.2077	118.7875	249.8144	106.6543	120.6237

Before ending this section, it should be noted here that the process of sorting has been mentioned a number of times during the example. It is suggested that sorting should be included at the mentioned places by default even if it is not required, as sorting functions are available in various programming tools.

Also in the above example, since the calculations have been done for only two iterations the objective function value of the worst frog always improves in the memetic evolution. However, with increase in the number of iterations, the difference between the best frog and the worst frog in the submemeplex may not be enough to obtain an improvement in the objective function value of the worst frog. In such a case, the step size will have to be again evaluated with the best frog in the swamp, and if that also does not help, then a random frog will be required to be generated.

The memeplexes are recombined to form the matrix 'X', and the process will be repeated until the convergence is met. Fig 7.9 shows the variation of 'P_X' with respect to the various iterations. The output in the figure, has the following parameters: number of frogs – 6, number of memeplexes – 2, number of frogs in one memeplex – 3, number of frogs in a submemeplex – 2, number of iterations – 200 and memetic iterations for one memeplex – 5.

The minimum objective function value at the end of 200 iterations was found to be 0.4726.

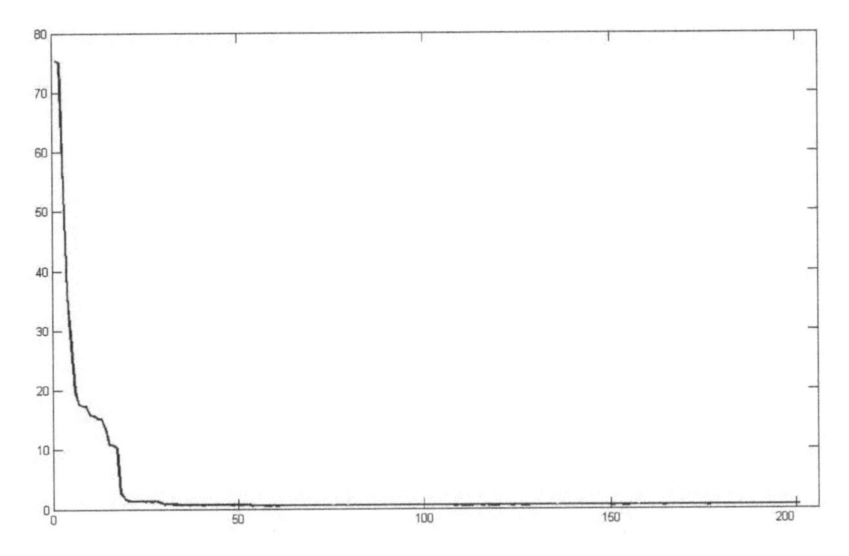

Figure 7.9: Variation of P_X through Iterations in SFLA

8 Grey Wolf Optimizer

8.1 INTRODUCTION

The Grey Wolf Optimizer (GWO) is a promising optimization technique. Similar to other EOAs, it also puts the base of its search process on a biological event. GWO is inspired from the hunting methods of wolves. This method was introduced in [99], and has been cited by a large number of researchers.

The wolves hunt in a pack. It is the coordination amongst the pack members that results in the hunt being successful. The hierarchy in the pack plays an important role, in which some of the wolves assume the roles of the leader. The pack is lead by two 'alpha' wolves, one male and another female. The alpha wolves exercise ultimate authority over the pack and are the undisputed leaders. The second level of wolves in the hierarchy of the pack are called 'beta'. They can be either male or female. The beta wolves are the best candidates to replace the alpha wolves, in case of death or old age of the later. The wolves in the next level of hierarchy are termed as 'delta'. The wolves in this category are usually scouts (to defend boundaries), sentinels, elders, hunters and caretakers. The lowest level of wolves are termed as 'omega'.

The hunting process of the grey wolves is classified into different stages as:

- Tracking, Chasing and Approaching the prey
- Pursue, encircle and harass the prey
- Attack

In [99], these steps are shown through some figures. Fig 8.1 tries to show the position of wolves during various stages of the hunt. The alpha wolf is assumed to be closest to the prey followed by beta and delta wolves. The wolves approach the prey and then encircle the prey before attacking it. The prey is represented with a dashed line circle to represent that in real life, the prey is visible however, in the optimization technique the prey is not visible (position not known) and it is assumed that the alpha wolf along with beta and delta wovles have a better idea about the location of the prey. The wolf nearest to the prey is assumed to be alpha, the next nearest wolf is assumed to be beta and the third nearest wolf to the prey is the delta. The remaining wolves are termed as omega. These wolves are represented with an 'A', 'B', 'D' and 'O' within the circles, while the prey is given by 'P' in a dashed circle.

In GWO, random wolves are generated for hunting the prey i.e., the optimal solution. The solution having the best optimization function value in the pack is termed as alpha wolf, the second best is termed as beta wolf and the third best is termed as delta wolf. The remaining solutions are termed as omega wolves. The next positions of the wolves in the pack shall be found by taking into consideration the effect of influence of the leader wolves. The leader wolves shall also move based on these same evaluations. The next iteration shall have new leader wolves and the process

DOI: 10.1201/b22647-8

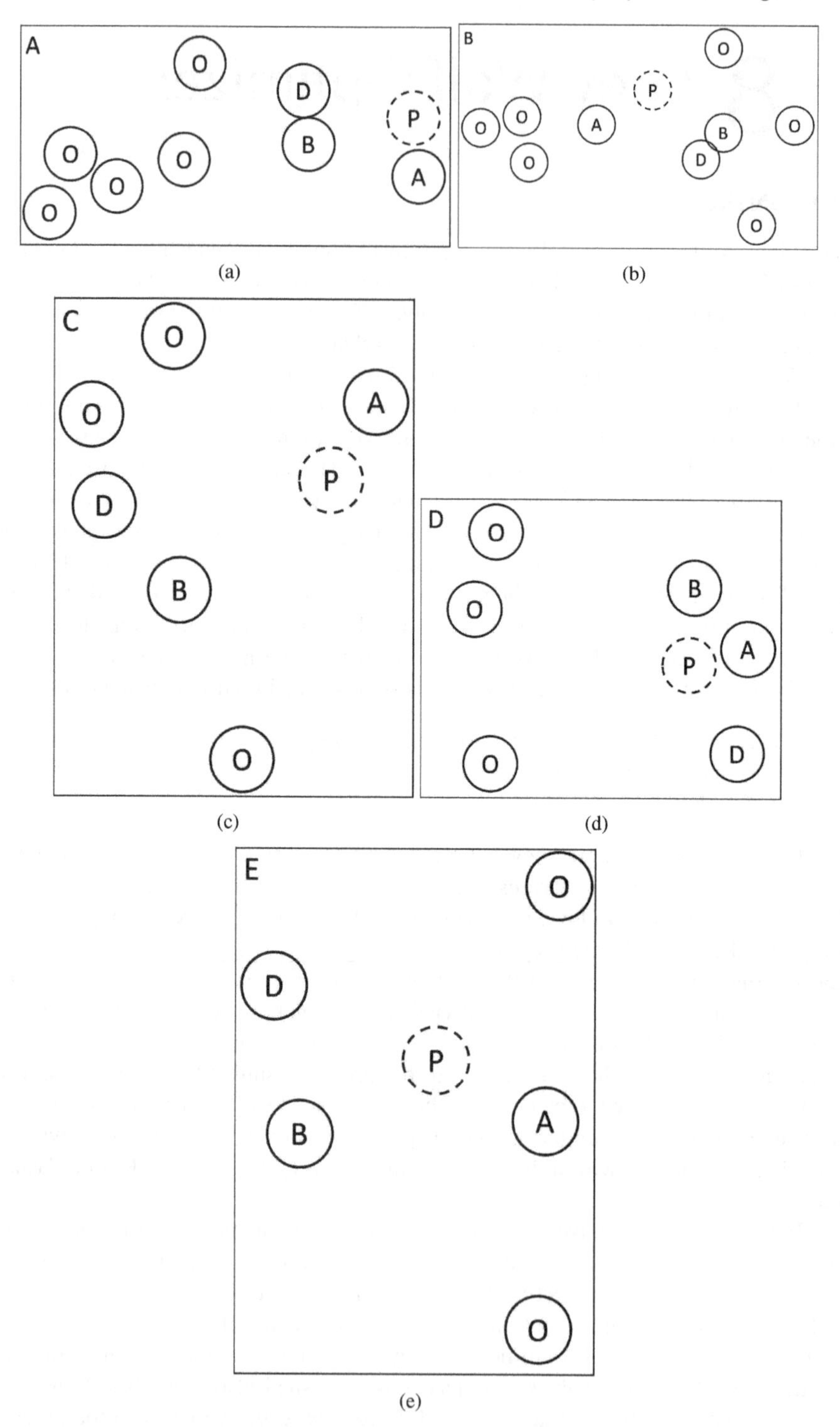

Figure 8.1: Hunting Steps for Grey Wolf Pack

will continue until the stopping criterion is met. The whole process is explained in later sections.

Before going towards the detailed analysis and step-by-step solution of GWO, let us discuss some critical points which make the optimization process a little bit different from the real hunt process of the grey wolves:

- The number of alpha wolves as mentioned in the reference is two, the number of beta and delta wolves are more, but in the optimization process they are restricted to one each.
- The alpha, beta and delta wolves cannot change ranks in a real pack, whereas there are new alpha, beta and delta wolves in each iteration.
- In the real hunt the prey is clearly visible, however, in the optimization technique/prey is not known. Also this prey can move in the real world, but it is stationary and needs to be searched in the optimization process.
- In the optimization process, the wolves move according to the movement of the leader wolves, but in reality these wolves can also see the prey and make their own moves.
- Randomness in the optimization process does not map to any movement or habit of the wolves during the hunt. The published work also does not give any specific reasoning for the random numbers.
- Ideally speaking no two wolves can occupy the same position physically, but in the optimization process two wolves can occupy the same position if their input variable values are the same.
- The pack size is usually restricted to a low number, however, in the optimization technique we can have a large number of wolves.
- In a physical hunt alpha, beta and delta would be near each other, however, this may not be the case in the optimization search process, and these solutions may occupy distant positions in the search space.

After studying the GWO optimization process, it is also suggested that the readers try and experiment with the following:

- Create two alpha wolves, multiple beta and delta wolves in the pack and observe the change in the performance of the GWO.
- Update the positions of the leader wolves dynamically, instead of waiting until the end of the iteration.
- Stopping crtieria can also be taken as: (i) alpha wolf stops moving or (ii) the whole pack gathers in the same position.

GWO has got some really good qualities like the equations used are quite simple and the matrix dimensions of various variables remain the same. Also, GWO has one advantage that the code is available at the website mentioned in [99].

8.2 TERMINOLOGY

This section presents some of the terms which relates the wolf hunting and pack behavior with the optimization process. Some of the terminologies are self assumed based on the working of the optimization method:

1. **Wolf**: A candidate solution to the optimization problem is termed as a wolf. It will be a row matrix whose size should be equal to the number of input variables.
2. **Pack**: A collection of wolves/candidate solutions is a pack.
3. **Alpha**: It is a solution/wolf which leads the pack during the hunt. In other words it is the best solution within the pack.
4. **Beta**: It is a solution which has the second best/optimal objective function value. It is also utilized for guiding the pack towards optimal solution.
5. **Delta**: It is the third best/optimal solution. It is used in combination with the Alpha and Beta wolf in guiding the pack towards the hunt/optimal solution.
6. **Omega**: The wolves/solutions in the pack other than alpha, beta and delta, are considered as omega wolves/solutions. They move towards the hunt with the help of alpha, beta and delta solutions.
7. **Social Hierarchy**: The importance allocated to a wolf in a pack. The wolves which are in the higher order of hierarchy, lead the pack in the hunt. In optimization process, the solutions having better objective function value are placed higher up in the heirarchy and are required to lead the optimization process towards the optimal solution.
8. **Encircling of Prey**: In the Grey Wolf hunting strategy, the pack first encircles its prey before going for the kill. In the optimization process, this activity is mathematically modeled through certain equations mentioned in the later sections.
9. **Hunting**: In reality the Alpha wolf usually guides/leads the hunt. The Beta and Delta wolves, occasionally participate in the hunt [99]. In mathematical simulation, it is however, assumed that all three participate and guide the other wolves in the hunt. This process is realized through some equations mentioned in the later sections.
10. **Attacking the Prey**: The wolves attack the prey once it stops moving. This is associated with exploitation in the search process. The attack on the prey is achieved by controlling some of the constants and their variation throughout the search process. Attacking the prey could have also been associated with the stopping criterion, but not considered in the reference introducing the method.

8.3 FUNDAMENTAL CONCEPT

The GWO process of optimization is derived from the hunting strategies of grey wolves also called canis lupus. These wolves live in a pack and hunt in a pack. Over a period of centuries these wolves have evolved their hunting strategies and have optimized it.

These wolves usually hunt in a pack having an average size of 5 to 12 wolves [99]. The grey wolf pack follows a strict heirarchy within itself. The wolves that lead the pack in the hunt are called alphas, one male and one female. The alphas control the pack and are responsible for the overall behavior of the pack.

The beta wolves form the second level of heirarchy in the pack. They facilitate the alpha wolves with decision making and other activities. Beta wolf can be a male or a female. The beta wolf would eventually replace the alpha wolf at sometime when the alpha dies or becomes old.

Another category in the wolf pack are the deltas, which are usually the scouts, sentinels, elders or caretakers. This category is placed below the beta and can still hold some sway within the pack. They patrol the territorial boundaries of the pack and look out for the enemies. They also take care of other wolves in the pack like the pups, elders, weak, ill, etc. The deltas will also play a part in the hunt. The last in the heirarchy are the omegas. These wolves do not have any rights within the pack and have to follow others.

The hunting process is basically divided into 3 parts [99]: (i) Track, chase and approach the target (ii) Pursue, encircle, and harass the prey and (iii) attack the prey.

The Grey Wolf hunting method is reproduced for optimization technique formulation in GWO. The pack represents a group/collection of solutions. The best solution is termed the alpha (α) wolf. The second best and the third best solutions are termed beta (β) and delta (δ) wolves, respectively. The remaining wolves in the pack are termed omega (ω) wolves. The optimization process, i.e., the hunt is guided by $\alpha, \beta and \delta$ wolves. The remaining of the pack consisting of the ω wolves will follow these three wolves. This is the social heirarchy context of the hunting process.

The next step in the hunt process is the encircling of the prey. For encircling of the prey, the mathematical model proposed consists of following two equations 8.1 and 8.2:

$$\vec{D} = \left| \vec{C} \cdot \vec{X}_p(t) - \vec{X}(t) \right| \tag{8.1}$$

$$\vec{X}(t+1) = \vec{X}_p(t) - \vec{A} \cdot \vec{D} \tag{8.2}$$

where:

$\vec{D}$: an intermediate set of values. It is not given any name for reference. It is further divided into three parts later on. It should be noted that $\vec{D}$ should be an absolute value and thus cannot have negative values.
$\vec{A}$ and $\vec{C}$: coefficients vectors. Their calculation is given in the next part.
t : current iteration
$\vec{X}_p(t)$: represents the position of the prey in the search space. It is a vector quantity and it should have all the input vector values.
$\vec{X}(t)$: represents the position of the grey wolf in consideration

The coefficient vectors $\vec{A}$ and $\vec{C}$ are defined in equations 8.3 and 8.4.

$$\vec{A} = 2\vec{a} \cdot \vec{r}_1 - \vec{a} \tag{8.3}$$

$$\vec{C} = 2 \cdot \vec{r}_2 \tag{8.4}$$

where:

$\vec{A}$ and $\vec{C}$: coefficient vectors, each having a dimension similar to a wolf/solution.

$\vec{a}$: it is linearly decreased from 2 to 0 over the course of the search process.
$\vec{r}_1$ and $\vec{r}_2$: random numbers from 0 to 1.

In the next phase, comes the hunting process. The omegas are guided towards the prey by the alpha, beta and delta wolves. Eqs 8.5, 8.6 and 8.7 represent the effect of the leader wolves on omega wolves (and on themselves). These equations are a modification of Eq 8.1 into three different forms for Alpha, Beta and Delta wolves.

$$\vec{D}_\alpha = \left|\vec{C}_1 \cdot \vec{X}_\alpha - \vec{X}\right| \tag{8.5}$$

$$\vec{D}_\beta = \left|\vec{C}_2 \cdot \vec{X}_\beta - \vec{X}\right| \tag{8.6}$$

$$\vec{D}_\delta = \left|\vec{C}_3 \cdot \vec{X}_\delta - \vec{X}\right| \tag{8.7}$$

On a similar note, to accomodate the three leader wolves for Eq 8.2, it is modified into Eqs 8.8, 8.9 and 8.10.

$$\vec{X}_1 = \vec{X}_\alpha - \vec{A}_1 \cdot \left(\vec{D}_\alpha\right) \tag{8.8}$$

$$\vec{X}_2 = \vec{X}_\beta - \vec{A}_2 \cdot \left(\vec{D}_\beta\right) \tag{8.9}$$

$$\vec{X}_3 = \vec{X}_\delta - \vec{A}_3 \cdot \left(\vec{D}_\delta\right) \tag{8.10}$$

The final effect of the leader wolves, i.e., the alpha, the beta and the delta, on the other wolves (and on themselves) is given through the Eq 8.11.

$$\vec{X}(t+1) = \frac{\vec{X}_1 + \vec{X}_2 + \vec{X}_3}{3} \tag{8.11}$$

where $\vec{X}(t+1)$: new position of the wolf under consideration.

These steps are repeated for all the wolves including the Alpha, Beta and the Delta. It should be kept in mind to check whether any of the input variable value does not exceed the permissible limits. The objective function value for each wolf is evaluated for their new positions. A new Alpha, Beta and Delta are found based on the objective function values.

The iteration ends here. These iterations are repeated until the stopping criterion is satisfied. The stopping criterion would usually be decided according to certain number of iterations. One can keep track of the Alpha wolf positions during the execution of the optimization process to observe the movement of the search process.

8.4 ALGORITHM AND PSEUDOCODE

The algorithm for GWO is given in Algorithm 8.1.

Algorithm 8.1 Grey Wolf Optimizer

```
START
Input Required Data
for 1 to Pack Size do
   Generate Wolf
   for 1 to Number of Variables do
      if Variable > Max Limit or Variable < Min Limit then
         Marks = Limit
      end if
   end for
end for
for 1 to Pack Size do
   Obtain Objective Function Value
end for
for 1 to Maximum Number of Iterations do
   Find Alpha, Beta and Delta Wolves.
   for 1 to Pack Size do
      for 1 to Number of Variables do
         Find A1, C1, Dα and X1
         Find A2, C2, Dβ and X2
         Find A3, C3, Dδ and X3
         Find X_New
      end for
   end for
   for 1 to Pack Size do
      for 1 to Number of Variables do
         if Variable > Max Limit or Variable < Min Limit then
            Marks = Limit
         end if
      end for
   end for
   for 1 to Pack Size do
      Obtain Objective Function Value
   end for
end for
Display Result
STOP
```

In the first part of the algorithm, the wolves are randomly generated. The input variable limits for these wolves/solutions should be checked. After generating the wolves, the objective function value of each wolf is evaluated.

As the initial steps are completed, the algorithm is ready to start the search process. The optimization technique shall repeat for maximum number of iterations as shown in the Algorithm. In the first step of the iteration, we mark the three best wolves in the pack as Alpha, Beta and Delta, respectively.

A nested loop is then employed, in which the outer loop repeats for the pack size, i.e., the number of wolves and the inner loop iterates for number of variables. Inside this loop, A_1, C_1, D_α and X_1 are found using Eqs 8.3, 8.4, 8.5 and 8.8, respectively. Similarly, A_2, C_2, D_β and X_2 are calculated using Eqs 8.3, 8.4, 8.6 and 8.9. In the third step of the loop, A_3, C_3, D_δ and X_3 are calculated using Eqs 8.3, 8.4, 8.7 and 8.10. As a last part of the iteration, X_{New} is evaluated through Eq 8.11.

Once the new positions of the wolves have been finalized, the input variable limits are checked and the objective function value for each wolf is obtained. This ends one iteration of the GWO. In the next iteration, the pack will again be sorted according the objective function value of the wolves and the leader wolves shall be marked. The loop shall continue until the stopping criterion is met.

8.5 FLOWCHART

The main flowchart of GWO is represented in Fig 8.2. The optimization technique starts with the random generation of wolf positions. It is assumed that the required input values, like the limits of the variables, etc. are read in the beginning. As a next step the objective function values of these wolves are evaluated. The position of the wolf will decide the heirarchy of the wolves.

The best three wolves in the pack are nominated as Alpha, Beta and Delta wolves, in the same order. Thus Alpha wolf will have the best solution in the pack, while Beta would be the second best solution and Delta will follow the Beta solution having the third best solution. The nomination of the Alpha, Beta and Delta wolves will also mark the beginning of the iterations in the optimization process.

The next part of the optimization process has a lot of equations to be solved and is the heart of the process. This process is separately presented in Fig 8.3. To obtain the new positions of the wolves, it is necessary to find the influence of the Alpha, Beta and Delta wolves. This influence is obtained through X_1,X_2 and X_3. The calculation of X_1,X_2 and X_3, further requires the values of A, C and D for each of them. Thus to obtain X_1, three more variables are to be calculated, and the same is applicable for X_2 and X_3. These X_1,X_2 and X_3 are to be calculated for each input variable of each wolf.

If we have 30 wolves with 5 input variables, then X_1,X_2 and X_3 will have to be evaluated for 150 times (30*5). Also each X will be accompanied by evaluation of 3 more variables, which means that for the above case, 600 equations (150*4) will have to be solved in one iteration, to find the new positions of the wolves. These calculations exclude the X_{New} equation, which will further add 150 calculations. The point of discussing the number of calculations is to indicate the time that will be required for executing these equations.

After evaluation of X_1,X_2 and X_3, let's return back to Fig 8.2. Now that we have got X_1,X_2 and X_3, we need to find X_{New}. The values of X_{New} will give the new positions of the wolves. Thus as mentioned before, X_{New} has to be evaluated for each input variable of each wolf.

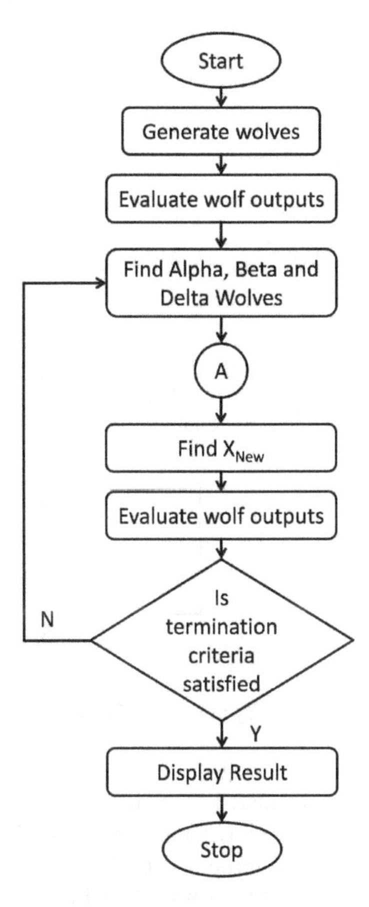

Figure 8.2: Flowchart for GWO

After obtaining the new wolf positions, we need to again evaulate the objective function values for each new wolf position. After evaluating the wolf outputs, we would check the termination criterion. As mentioned in the references, the termination criterion would be based on number of iterations. However, taking a cue from PSO, convergence can always be considered for terminating the optimization process. If the termination criterion is satisfied, the results are displayed and the process stops. In case the stopping criterion is not met, then Alpha, Beta and Delta wolves are again nominated from the pack and the loop is repeated.

8.6 EXAMPLE

The example section is introduced in every chapter to solve the standard Griewank optimization problem through the optimization technique studied, up to two iterations. Eqs 2.5 and 2.9 can be referred for understanding the Griewank problem and

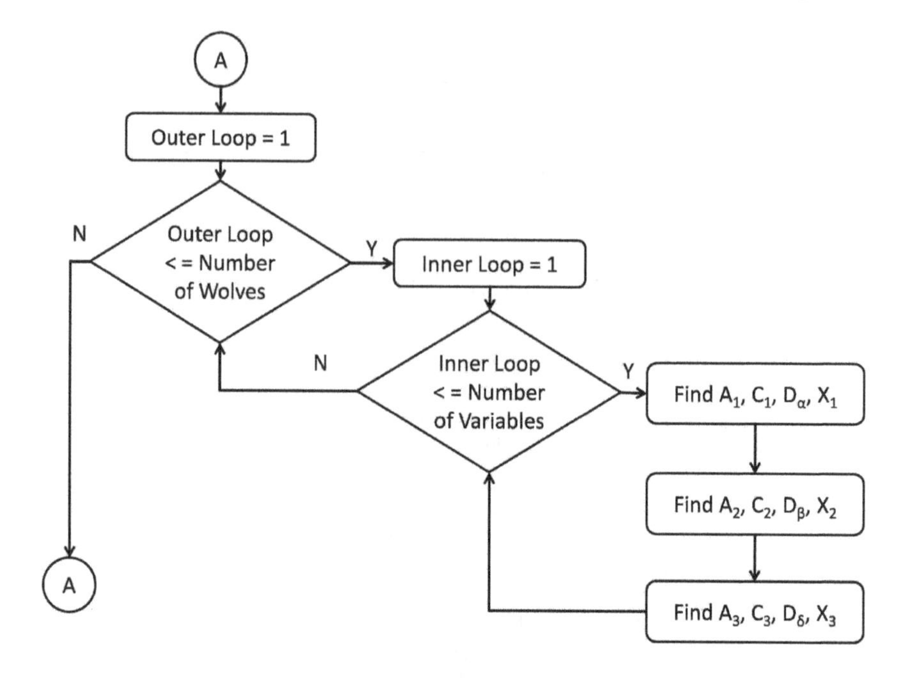

Figure 8.3: X_1, X_2 and X_3 in GWO

how its objective function can be calculated. A wolf is similar to a particle in PSO and can be represented as given in Table 5.1. The evaluation of the objective function value for a wolf can also be understood through Eq 5.9.

The wolf pack would be similar to a swarm in PSO. Let the pack be represented by Table 8.1.

Table 8.1
Wolf Pack

	Variable 1	Variable 2	Variable 3	Variable 4	Variable 5
Wolf 1	141.8811	−104.656	−397.606	211.7688	561.0059
Wolf 2	−16.0475	−277.517	74.94419	247.9742	391.221
Wolf 3	503.4618	202.9188	51.23069	356.5671	321.7123
Wolf 4	−80.9499	−329.007	−180.324	77.46379	397.8965
Wolf 5	471.3352	−152.64	86.91336	386.1857	−460.428
Wolf 6	−556.071	286.2077	118.7875	249.8144	106.6543

Objective function values are found for the wolves in the pack and are given in Table 8.2. The best/lowest three objective function values are shown in bold. These wolves will be selected as alpha, beta and delta wolves for the remaining part of the iteration.

Table 8.2
Objective Function Values for the Pack

	Objective Function Value
Wolf 1	138.1064
Wolf 2	**75.35377**
Wolf 3	132.9249
Wolf 4	**78.76276**
Wolf 5	154.5424
Wolf 6	**120.6237**

Table 8.3 shows the alpha, beta and delta wolves. The wolf 2 from the pack has the lowest objective function value and is termed the alpha. The second lowest objective function is that of wolf 4 and it is termed beta. The wolf 6 is made the delta wolf with the third lowest objective function value.

Table 8.3
Alpha, Beta and Delta Wolves

Wolf	Variable 1	Variable 2	Variable 3	Variable 4	Variable 5	Obj. Funct. Value
Alpha	−16.0475	−277.517	74.94419	247.9742	391.221	75.35377
Beta	−80.9499	−329.007	−180.324	77.46379	397.8965	78.76276
Delta	−556.071	286.2077	118.7875	249.8144	106.6543	120.6237

In the next step, the coefficient vectors A and C are to be obtained, for which we need random numbers. Even though these random numbers can be generated at run time, for the purpose of understanding the calculations, the random numbers are generated and tabularized in Table 8.4. These random numbers shall be used for calculating 'A_1' and 'C_1' corresponding to the effect of Alpha wolf. While executing

Table 8.4
Set of Random Numbers for X_1 in Iteration 1

Random Numbers R_1					Random Numbers R_2				
0.2437	0.3075	0.9185	0.4063	0.9165	0.1692	0.0275	0.2067	0.0302	0.0256
0.2726	0.9735	0.7196	0.3996	0.8061	0.0752	0.2126	0.8837	0.8665	0.8848
0.2074	0.1619	0.9812	0.3159	0.4218	0.7285	0.0623	0.8043	0.8189	0.297
0.4641	0.9814	0.9606	0.1252	0.1284	0.7771	0.1934	0.1933	0.776	0.0259
0.9102	0.1229	0.7075	0.3607	0.2746	0.3256	0.3388	0.0038	0.2423	0.1516
0.1118	0.2903	0.6623	0.8663	0.2627	0.5339	0.3909	0.1157	0.0671	0.6928

the GWO, it was noted that if the random numbers are kept the same, i.e., all the random numbers will be the same in Table 8.4, then the objective function value of Alpha wolves do not decrease over the iterations, instead a rise is observed in the objective function values. Thus the random numbers, play an important role in the optimization process.

From the random numbers given in Table 8.4, the values of 'A_1' are obtained using Eq 8.3. However, before obtaining the values of 'A_1', 'a' has to be found for the iteration. Assuming, that the optimization technique shall run for 200 iterations, 'a' for the first iteration can be found as:

$$a = 2 - \left(iteration\ number * \frac{2}{total\ iterations}\right) = 2 - \left(1 * \frac{2}{200}\right) = 1.99$$

The value of 'a' is supposed to reduce linearly from 2 to 0 throughout the search process. The above method of calculation can be employed to obtain this linearly reducing values of 'a'. The '2' in the above calculation represents the maximum value of 'a', '1' gives the iteration number and '200' gives the total number of iterations.

According to Eq 8.3, A_1 can be found as:

$$A1 = (2 * a * r_1) - a = (2 * 1.99 * 0.2437) - 1.99 = -1.020074$$

Similarly 'A_1' can be found for the remaining input variable values of the wolf as well as for the whole pack. In the above equation, '2' and '1.99' will remain as it is, only the random numbers shall change according to the values of 'r_1' in Table 8.4. The values of 'A_1' for the whole pack have been calculated and tabularized in Table 8.5.

Table 8.5

A_1 Values for Iteration 1

Variable 1	Variable 2	Variable 3	Variable 4	Variable 5
−1.02007	-0.76633	1.665451	−0.37312	1.65775
−0.90496	1.884381	0.873892	−0.39945	1.218106
−1.16472	−1.34551	1.915195	−0.73286	−0.31116
−0.14278	1.915985	1.833104	−1.49178	−1.47914
1.632628	−1.50085	0.825832	−0.55428	−0.89725
−1.54511	−0.83447	0.645885	1.457693	−0.94456

The calculation of the other coefficient vector 'C_1' is very simple, and is given by Eq 8.4. To find the values of 'C_1', the values of random number 'r_2' is doubled. Table 8.6 gives the values of 'C_1' for the first iteration.

Table 8.6
C_1 Values for Iteration 1

Variable 1	Variable 2	Variable 3	Variable 4	Variable 5
0.338312	0.054919	0.413498	0.060434	0.051179
0.150431	0.425122	1.767318	1.733048	1.769506
1.457057	0.124647	1.608649	1.637753	0.594061
1.554135	0.386782	0.386666	1.551989	0.051789
0.651191	0.67754	0.007523	0.484721	0.303159
1.067702	0.781753	0.231364	0.134246	1.385531

In the next step, D_α is to be found. The values of D_α can be obtained through Eq 8.5. A sample calculation for the same if given below:

$$\begin{aligned}\vec{D}_\alpha &= \left|\vec{C}_1 \cdot \vec{X}_\alpha - \vec{X}\right| \\ &= |0.3383 * (-16.0475) - 141.8811| \\ &= |-5.4289 - 141.8811| \\ &= 147.3102\end{aligned}$$

The calculation of D_α is enclosed in a mod sign, that means that the output should always be the magnitude and will always remain positive. The values taken are as follows:

0.3383 : Table 8.6
−16.0475 : Table 8.3, first value for Alpha wolf.
141.8811 : Table 8.1, first value of Wolf 1

To calculate the remaining values of D_α, the values to be taken from Tables 8.6 and 8.1 will move correspondingly with the change in row and column, i.e., if the input variable changes then the column should change for the value to be selected in both the tables, and if the wolf changes then the row should change in both the tables. However, for Table 8.3, the column will only change according to the change in input variable, the row will not change.

Let us take another example, by obtaining the value of D_α for 3rd wolf and 2nd input variable.

$$D_\alpha = |0.1264 * (-277.517) - 202.9188| = 237.9969$$

Note: The values may vary a little after decimals, as the numbers in various tables have been rounded to 4 decimal places.

In the above calculation,

0.1264 : 3rd row and 2nd column value from Table 8.6
−277.517 : 1st row and 2nd column (input variable) value from Table 8.3.

202.9188 : 3rd row and 2nd column (input variable) value from Table 8.1

Similarly, the remaining values for D_α can be obtained. These values are tabularized in Table 8.7.

Table 8.7
D_α Values for Iteration 1

Variable 1	Variable 2	Variable 3	Variable 4	Variable 5
147.3102	89.41493	428.5955	196.7828	540.9836
13.63343	159.5383	57.50605	181.7769	301.0468
526.8439	237.5104	69.32823	49.55332	89.30309
56.00998	221.6684	209.3021	307.3895	377.6355
481.7852	35.38857	86.34955	265.9874	579.0298
538.9368	503.1573	101.4481	216.5249	435.3947

In the last part of calculation, related to the influence of Alpha wolf, we find X_1. Eq 8.8 is used for obtaining the values of X_1 as shown below:

$$\begin{aligned} X_1 &= X_\alpha - A_1 * D_\alpha \\ &= -16.0475 - (-1.02) * 147.3102 \\ &= -16.0475 + 150.2564 \\ &= 134.2089 \end{aligned}$$

where the values are taken from

-16.0475 : Table 8.3, first value.
-1.02 : Table 8.5, first value.
147.3102 : Table 8.7, first value.

The remaining values of X_1, are calculated in a similar way. The rows and columns shall change in correspondence for Tables 8.5 and 8.7, however, for Table 8.3, the columns will only change, without change in row. Another calculation is presented below for further improvement in understanding the values involved in obtaining X_1. Let us find the value for 4th wolf and 4th input variable.

$$X_1 = 247.9742 - (-0.5543) * 265.9874 = 395.411$$

In the above calculation,

247.9742 : 1st row and 4th column value from Table 8.3
-0.5543 : 4th row and 4th column value from Table 8.5.
265.9874 : 4th row and 4th column value from Table 8.7

Similarly, the remaining values for X_1 can be obtained. These values are tabularized in Table 8.8.

Table 8.8
X_1 Values for Iteration 1

Variable 1	Variable 2	Variable 3	Variable 4	Variable 5
134.2194	−208.995	−638.861	321.3984	−505.595
−3.7098	−578.148	24.69011	320.5851	24.51408
597.5764	42.05631	−57.8329	284.2899	419.0082
−8.05039	−702.23	−308.728	706.5311	949.7978
−802.623	−224.404	3.633998	395.4052	910.7549
816.6697	142.3511	9.420388	−67.6527	802.4794

One iteration in GWO is divided into three similar parts, one with Alpha wolf as influencer, second with Beta wolf as influencer and the last with Delta wolf as influencer. All the three parts of the iteration are similar except that the row from which data is used from Table 8.3 changes in each part. Thus for this part of the iteration, in which Beta wolf is the influencer, we would be using the data from second row of Table 8.3.

For the second part of the iteration, first the random numbers are generated and presented in Table 8.9. The random numbers are divided into r_1 and r_2.

Table 8.9
Set of Random Numbers for X_2 in Iteration 1

Random Numbers R_1					Random Numbers R_2				
0.7423	0.2162	0.3101	0.4092	0.5517	0.9568	0.3534	0.0073	0.3874	0.0644
0.3066	0.1383	0.9910	0.5010	0.8435	0.7429	0.7122	0.2560	0.3528	0.7337
0.0199	0.3992	0.8413	0.7619	0.9175	0.8712	0.5396	0.6201	0.8542	0.5618
0.8036	0.2939	0.9487	0.9865	0.4812	0.1503	0.3297	0.1095	0.4036	0.4363
0.0207	0.0525	0.8396	0.4977	0.0685	0.3229	0.4594	0.1504	0.8173	0.5911
0.8143	0.4068	0.3357	0.2526	0.3308	0.2260	0.5438	0.7720	0.2774	0.1660

Tables 8.10 and 8.11 give the values of 'A_2' and 'C_2', which are calculated based on Eqs 8.3 and 8.4, respectively. Since the iteration has not changed, the value of 'a' remains the same at 1.99.

In the next step, D_β is obtained using Eq 8.6. The result is tabularized in Table 8.12. The influence of Beta wolf given by X_2 is obtained through Eq 8.9, and tabularized in Table 8.13.

In the last part of the iteration the influence due to the Delta wolf is evaluated through the same process. In this part of the iteration, the third row from Table 8.3 shall be used. Let us list the tables for this part of the iteration.

Table 8.14 gives the random numbers corresponding to r_1 and r_2. Tables 8.15 and 8.16 gives the values of 'A_3' and 'C_3', obtained through Eqs 8.3 and 8.4, respectively.

Table 8.10
A_2 Values for Iteration 1

Variable 1	Variable 2	Variable 3	Variable 4	Variable 5
0.964177	−1.12936	−0.75574	−0.3614	0.205658
−0.7699	−1.43966	1.954259	0.004035	1.366969
−1.91067	−0.40117	1.358212	1.042364	1.661465
1.208376	−0.82032	1.785899	1.936131	−0.07474
−1.90747	−1.78089	1.351522	−0.00902	−1.71736
1.2511	−0.37104	−0.65379	−0.98466	−0.67335

Table 8.11
C_2 Values for Iteration 1

Variable 1	Variable 2	Variable 3	Variable 4	Variable 5
1.913541	0.706732	0.014573	0.77489	0.128743
1.485807	1.424456	0.511906	0.705557	1.467312
1.742322	1.079172	1.240216	1.708399	1.123516
0.300676	0.659358	0.21899	0.807297	0.87259
0.645848	0.918727	0.300899	1.634546	1.182153
0.452091	1.087625	1.544009	0.554705	0.331934

Table 8.12
D_β Values for Iteration 1

Variable 1	Variable 2	Variable 3	Variable 4	Variable 5
296.7821	127.864	394.9783	151.7429	509.7793
104.2285	191.1394	167.253	193.3191	192.6175
644.5027	557.9739	274.8711	224.228	125.3306
56.61025	112.0736	140.8347	14.9275	50.69612
523.6166	149.6274	141.1726	259.5676	930.8022
519.474	644.0439	397.2089	206.8449	25.42115

Table 8.17 gives the values of D_δ found through Eq 8.7 and Table 8.18 gives the values of X_3 found through Eq 8.10.

After obtaining the values of X_1, X_2 and X_3, the new position of the wolves can be found through Eq 8.11. It simply states that the new wolf position will be the average of X_1,X_2 and X_3. All the corresponding elements of matrices X_1,X_2 and X_3 are added and the sum is divided by 3. An example calculation for wolf 1 and input variable 1 is shown below:

Table 8.13
X_2 Values for Iteration 1

Variable 1	Variable 2	Variable 3	Variable 4	Variable 5
−367.1	−184.603	118.1784	132.3036	293.0562
−0.70429	−53.8306	−507.179	76.68379	134.5943
1150.484	−105.164	−553.657	−156.263	189.664
−149.356	−237.07	−431.84	48.5622	401.6857
917.8325	−62.5377	−371.122	79.80419	1996.416
−730.864	−90.0378	79.36912	281.1364	415.0139

Table 8.14
Set of Random Numbers for X_3 in Iteration 1

Random Numbers R_1					Random Numbers R_2				
0.9701	0.2864	0.6087	0.8101	0.3133	0.3686	0.3887	0.8122	0.6163	0.5571
0.1519	0.5893	0.2552	0.4305	0.8789	0.9990	0.9628	0.8059	0.5397	0.0289
0.7059	0.1175	0.5637	0.5787	0.6398	0.3091	0.7454	0.2487	0.6248	0.0264
0.2978	0.2602	0.7363	0.3898	0.1297	0.3329	0.8080	0.3179	0.4274	0.4926
0.1779	0.7811	0.8588	0.4178	0.0156	0.5001	0.4449	0.4744	0.9216	0.1018
0.6932	0.2812	0.2352	0.7054	0.5382	0.7427	0.3202	0.1988	0.1817	0.6909

Table 8.15
A_3 Values for Iteration 1

Variable 1	Variable 2	Variable 3	Variable 4	Variable 5
1.8712	−0.8501	0.4328	1.2343	−0.7432
−1.3853	0.3553	−0.9741	−0.2767	1.5080
0.8195	−1.5224	0.2536	0.3132	0.5564
−0.8046	−0.9544	0.9404	−0.4388	−1.4738
−1.2819	1.1189	1.4280	−0.3273	−1.9281
0.7691	−0.8709	−1.0540	0.8173	0.1519

$$X_{New} = \frac{X_1 + X_2 + X_3}{3} = \frac{134.2194 + (-367.1) + (-1588.59)}{3} = -607.157$$

Another calculation is presented for 5th wolf and 4th input variable.

$$X_{New} = \frac{X_1 + X_2 + X_3}{3} = \frac{(-67.6527) + 281.1364 + 119.8152}{3} = 111.0996$$

Table 8.16
C_3 Values for Iteration 1

Variable 1	Variable 2	Variable 3	Variable 4	Variable 5
0.7372	0.7773	1.6245	1.2326	1.1143
1.9980	1.9255	1.6118	1.0794	0.0578
0.6181	1.4908	0.4975	1.2496	0.0528
0.6657	1.6159	0.6357	0.8548	0.9852
1.0003	0.8898	0.9487	1.8433	0.2036
1.4855	0.6404	0.3976	0.3633	1.3819

Table 8.17
D_δ Values for Iteration 1

Variable 1	Variable 2	Variable 3	Variable 4	Variable 5
551.7984	327.1390	590.5722	96.1616	442.1647
1094.9605	828.6127	116.5135	21.6759	385.0575
847.1878	223.7670	7.8618	44.3905	316.0861
289.2526	791.5038	255.8408	136.0684	292.8210
1027.5696	407.3219	25.7817	74.2949	482.1432
269.9583	102.9346	71.5572	159.0516	40.7294

Table 8.18
X_3 Values for Iteration 1

Variable 1	Variable 2	Variable 3	Variable 4	Variable 5
−1588.5934	564.3132	−136.7880	131.1196	435.2723
960.8159	−8.1841	232.2850	255.8111	−474.0078
−1250.3572	626.8670	116.7941	235.9126	−69.2316
−323.3328	1041.6249	−121.7985	309.5191	538.2109
761.2101	−169.5535	81.9724	274.1341	1036.2529
−763.7023	375.8523	194.2062	119.8152	100.4673

The new positions for all the wolves is given in Table 8.19.

The values in bold, show the input variables which have exceeded their limits. They have to be brought within the limits of −600 to 600. The new wolf positions are given in Table 8.20. These new modified wolf positions shall serve as input to the next iteration of GWO.

The objective function values for these new modified wolf positions are found, which are given in Table 8.21.

Table 8.19
New Wolf Positions

	Variable 1	Variable 2	Variable 3	Variable 4	Variable 5
Wolf 1	**−607.1582**	56.9050	−219.1568	194.9405	74.2446
Wolf 2	318.8006	−213.3875	−83.4015	217.6933	−104.9665
Wolf 3	165.9010	187.9198	−164.8986	121.3130	179.8135
Wolf 4	−160.2465	34.1082	−287.4558	354.8708	**629.8981**
Wolf 5	292.1397	−152.1650	−95.1717	249.7812	**1314.4745**
Wolf 6	−225.9654	142.7219	94.3319	111.0996	439.3202

Table 8.20
New Modified Wolf Positions

	Variable 1	Variable 2	Variable 3	Variable 4	Variable 5
Wolf 1	−600	56.9050	−219.1568	194.9405	74.2446
Wolf 2	318.8006	−213.3875	−83.4015	217.6933	−104.9665
Wolf 3	165.9010	187.9198	−164.8986	121.3130	179.8135
Wolf 4	−160.2465	34.1082	−287.4558	354.8708	600
Wolf 5	292.1397	−152.1650	−95.1717	249.7812	600
Wolf 6	−225.9654	142.7219	94.3319	111.0996	439.3202

This ends the calculations for one iteration. Before we move to the next iteration, let us analyze the number of calculations to be performed in one iteration.

Let us also analyze if the matrices shown in the section need to be stored or not. If one input variable of a wolf for the new position is calculated at a go, then we need not store and form matrices for any of the above calculations except the new position. The flow of calculation would be:

$r_1, r_2 \rightarrow A_1, C_1 \rightarrow D_\alpha \rightarrow X_1 \rightarrow r_1, r_2 \rightarrow A_2, C_2 \rightarrow D_\beta \rightarrow X_2 \rightarrow r_1, r_2 \rightarrow A_3, C_3 \rightarrow D_\delta \rightarrow X_3 \rightarrow X_{New}$

Thus, only X_{New} is required to be stored, the remaining variables are temporary and change continuously. The matrices shown in this section are only for the sake of verification and understanding of the calculations and need not be stored.

In the new iteration, let us first find the Alpha, Beta and Delta wolves. The three lowest objective function values are given in bold in Table 8.21. These wolves shall serve as alpha, beta and delta in this iteration. In the last iteration, wolves 2, 4 and 6 were included as the leader wolves; however, in this iteration wolves 2, 3 and 6 are taken as leader wolves.

Table 8.22, in the second iteration gives the alpha, beta and delta wolves for this iteration. This table will now be used for all calculations related to alpha, beta and delta wolves in this iteration.

Table 8.21
Objective Function Values for Modified Wolf Positions

	Objective Function Value
Wolf 1	114.5814
Wolf 2	**54.1171**
Wolf 3	**35.2214**
Wolf 4	149.8596
Wolf 5	135.9913
Wolf 6	**72.3896**

Table 8.22
Alpha, Beta and Delta Wolves for Iteration 2

Wolf	Variable 1	Variable 2	Variable 3	Variable 4	Variable 5	Obj. Funct. Value
Alpha	165.9010	187.9198	−164.8986	121.3130	179.8135	35.2214
Beta	318.8006	−213.3875	−83.4015	217.6933	−104.9665	54.1171
Delta	−225.9654	142.7219	94.3319	111.0996	439.3202	72.3896

Table 8.23
Set of Random Numbers for X_1 in Iteration 2

Random Numbers R_1					Random Numbers R_2				
0.0550	0.8963	0.0753	0.0402	0.3679	0.6895	0.1162	0.5745	0.3392	0.4148
0.5742	0.7481	0.2337	0.9866	0.7745	0.2956	0.4363	0.6498	0.1654	0.5863
0.6436	0.0733	0.8658	0.2183	0.6617	0.8162	0.9031	0.9677	0.7966	0.4122
0.0905	0.1164	0.9111	0.1972	0.9268	0.5158	0.5806	0.7877	0.2644	0.2372
0.6308	0.8627	0.7183	0.3322	0.4366	0.9885	0.8049	0.0335	0.5648	0.2226
0.1641	0.5820	0.7974	0.1194	0.7986	0.7665	0.2679	0.8613	0.8723	0.1096

For this iteration, the value of ‘a’ can be calculated as:

$$a = 2 - \left(iteration\ number * \frac{2}{total\ iterations} \right) = 2 - \left(2 * \frac{2}{200} \right) = 1.98$$

Next we generate random numbers, for the obtaining X_1. The random numbers are given in Table 8.23.

The values of A_1 and C_1 are presented in Table 8.24.

The D_α values for iteration 2 are tabularized in Table 8.25.

Table 8.26 gives the values of X_1 for iteration 2.

The random numbers for obtaining X_2 is given in Table 8.27. The A_2 and C_2

Table 8.24
A_1 and C_1 for Iteration 2

A_1					C_1				
−1.7623	1.5693	−1.6817	−1.8209	−0.5230	1.3790	0.2323	1.1490	0.6785	0.8295
0.2939	0.9824	−1.0545	1.9268	1.0870	0.5913	0.8727	1.2996	0.3309	1.1725
0.5688	−1.6897	1.4487	−1.1154	0.6404	1.6323	1.8061	1.9355	1.5932	0.8245
−1.6217	−1.5191	1.6280	−1.1990	1.6902	1.0316	1.1611	1.5754	0.5287	0.4744
0.5182	1.4362	0.8643	−0.6644	−0.2510	1.9769	1.6098	0.0670	1.1295	0.4453
−1.3303	0.3248	1.1775	−1.5072	1.1826	1.5329	0.5358	1.7225	1.7446	0.2193

Table 8.25
D_α Values for Iteration 2

Variable 1	Variable 2	Variable 3	Variable 4	Variable 5
828.7768	13.2467	29.6958	112.6339	74.9162
220.7114	377.3811	130.8989	177.5564	315.8032
104.9016	151.4873	154.2551	71.9633	31.5574
331.3934	184.0919	27.6665	290.7313	514.6973
35.8378	454.6773	84.1215	112.7555	519.9372
480.2755	42.0417	378.3753	100.5487	399.8890

Table 8.26
X_1 Values for Iteration 2

Variable 1	Variable 2	Variable 3	Variable 4	Variable 5
1626.4664	167.1324	−114.9592	326.4054	218.9924
101.0236	−182.8250	−26.8645	−220.8108	−163.4703
106.2318	443.8822	−388.3611	201.5824	159.6055
703.3288	467.5691	−209.9397	469.8952	−690.1252
147.3316	−465.0979	−237.6043	196.2299	310.2962
804.7984	174.2667	−610.4441	272.8648	−293.0999

values are given in Table 8.28, whereas the D_β values are presented in Table 8.29. Lastly the X_2 values are tabularized in Table 8.30.

A similar set of tables are presented below for X_3. Table 8.31 for random numbers, Table 8.32 for A_3 and C_3, Table 8.33 for D_δ and Table 8.34 for X_3.

As a last step in iteration 2, we need to obtain the X_{New} values using Eq 8.11. These values are presented in Table 8.35. There were no input variables which crossed the specified limits. The last column gives the objective function value and the bold values indicate the Alpha, Beta and Delta wolves for the next iteration.

Table 8.27
Set of Random Numbers for X_2 in Iteration 2

Random Numbers R_1					Random Numbers R_2				
0.7687	0.9210	0.6876	0.2565	0.9334	0.5895	0.8698	0.7231	0.2395	0.5750
0.4709	0.0131	0.9953	0.5617	0.1919	0.4408	0.8429	0.2703	0.6620	0.6115
0.3917	0.1636	0.1783	0.3124	0.2378	0.9270	0.6419	0.7877	0.8491	0.7054
0.5487	0.6060	0.3591	0.7533	0.2309	0.1398	0.7524	0.4920	0.1277	0.1140
0.4427	0.7189	0.6740	0.5218	0.4549	0.5085	0.6713	0.4838	0.7602	0.7618
0.6698	0.1132	0.7176	0.8596	0.0879	0.6286	0.8531	0.1312	0.4149	0.4402

Table 8.28
A_2 and C_2 for Iteration 2

A_2					C_2				
1.0642	1.6672	0.7429	−0.9644	1.7163	1.1790	1.7396	1.4462	0.4790	1.1500
−0.1152	−1.9279	1.9616	0.2443	−1.2200	0.8816	1.6857	0.5406	1.3241	1.2230
−0.4288	−1.3320	−1.2741	−0.7430	−1.0383	1.8541	1.2838	1.5754	1.6983	1.4107
0.1927	0.4198	−0.5578	1.0032	−1.0655	0.2797	1.5048	0.9840	0.2554	0.2281
−0.2269	0.8667	0.6889	0.0862	−0.1787	1.0169	1.3425	0.9675	1.5205	1.5236
0.6722	−1.5319	0.8618	1.4239	−1.6319	1.2573	1.7062	0.2624	0.8298	0.8804

Table 8.29
D_β Values for Iteration 2

Variable 1	Variable 2	Variable 3	Variable 4	Variable 5
975.8666	428.1059	98.5445	90.6725	194.9535
37.7480	146.3296	38.3176	70.5519	23.4104
425.1866	461.8644	33.5107	248.3890	327.8898
249.4089	355.2071	205.3862	299.2773	623.9381
32.0559	134.3148	14.4780	81.2114	759.9270
626.7882	506.8109	116.2183	69.5330	531.7362

At the end of iteration 2, the wolves 2, 4 and 1 will be chosen as Alpha, Beta and Delta wolves, respectively for the next iteration. The least objective function value was obtained to be 14.5070 for wolf 2.

The program of GWO was executed for 200 times. The variation in the position of the Alpha is plotted for the execution in Fig 8.4. The objective function value at the end of 200 iterations was 0.0304. This method of showing the performance of an EOA is presented in many number of research papers and would be helpful for young researchers. The x-axis gives the number of iterations and the y-axis represents the

Table 8.30
X_2 Values for Iteration 2

Variable 1	Variable 2	Variable 3	Variable 4	Variable 5
−719.7242	−927.1253	−156.6071	305.1397	−439.5724
323.1486	68.7252	−158.5641	200.4603	−76.4065
501.1151	401.8274	−40.7056	402.2495	235.4879
270.7387	−362.4966	31.1678	−82.5546	559.8554
326.0732	−329.7937	−93.3748	210.6954	30.8158
−102.5346	562.9844	−183.5570	118.6819	762.7785

Table 8.31
Set of Random Numbers for X_3 in Iteration 2

Random Numbers R_1					Random Numbers R_2				
0.7256	0.4745	0.1820	0.1399	0.8806	0.7965	0.5604	0.1838	0.7010	0.7300
0.5982	0.8000	0.9730	0.5655	0.3465	0.8015	0.7467	0.0780	0.9019	0.2781
0.5625	0.7813	0.5413	0.7481	0.2685	0.1829	0.2303	0.1312	0.6674	0.0053
0.0185	0.9985	0.1629	0.2050	0.6924	0.3137	0.8868	0.7672	0.4463	0.7994
0.1149	0.1264	0.0442	0.5412	0.1289	0.1320	0.8259	0.8364	0.9082	0.5043
0.8717	0.6305	0.3452	0.9221	0.7325	0.9570	0.4581	0.3133	0.6313	0.6281

Table 8.32
A_3 and C_3 for Iteration 2

A_3					C_3				
0.8934	−0.1011	−1.2593	−1.4262	1.5071	1.5930	1.1208	0.3676	1.4020	1.4601
0.3888	1.1880	1.8729	0.2592	−0.6077	1.6029	1.4934	0.1560	1.8037	0.5561
0.2473	1.1140	0.1634	0.9823	−0.9168	0.3657	0.4607	0.2624	1.3347	0.0105
−1.9069	1.9739	−1.3348	−1.1682	0.7619	0.6275	1.7737	1.5344	0.8925	1.5988
−1.5252	−1.4793	−1.8050	0.1631	−1.4694	0.2641	1.6519	1.6728	1.8165	1.0087
1.4718	0.5168	−0.6131	1.6715	0.9208	1.9140	0.9163	0.6266	1.2627	1.2561

objective function value. It should also be observed that the Alpha objective function value of the graph temporarily increases in the graph. This may or may not happen for every execution. However, an increase in Alpha objective function value happens because the best Alpha position is not stored. Thus this can be added to the original GWO, for better performance. Also the Alpha, Beta and Delta wolves can be updated as soon as one of the wolf position appears to be a candidate position for the leader wolf position.

Table 8.33
D_δ Values for Iteration 2

Variable 1	Variable 2	Variable 3	Variable 4	Variable 5
240.0265	103.0583	253.8326	39.1798	567.1901
681.0073	426.5234	98.1133	17.2989	349.2792
248.5441	122.1737	189.6489	26.9763	175.1878
18.4592	219.0343	432.2014	255.7113	102.3822
351.8101	387.9227	252.9720	47.9689	156.8585
206.5389	11.9487	35.2271	29.1835	112.5259

Table 8.34
X_3 Values for Iteration 2

Variable 1	Variable 2	Variable 3	Variable 4	Variable 5
−440.4165	153.1414	413.9933	166.9766	−415.4682
−490.7394	−363.9923	−89.4249	106.6154	651.5872
−287.4345	6.6256	63.3468	84.6006	599.9380
−190.7653	−289.6389	671.2196	409.8187	361.3186
310.6087	716.5709	550.9530	103.2755	669.8062
−529.9566	136.5463	115.9304	62.3196	335.7098

Table 8.35
New Wolf Positions after Iteration 2

	Variable 1	Variable 2	Variable 3	Variable 4	Variable 5	Obj. Func. Value
Wolf 1	155.4419	−202.2838	47.4757	266.1739	−212.0161	**46.7821**
Wolf 2	−22.1891	−159.3640	−91.6178	28.7550	137.2368	**14.5070**
Wolf 3	106.6375	284.1117	−121.9066	229.4775	331.6771	68.3895
Wolf 4	261.1007	−61.5221	164.1492	265.7197	77.0163	**44.4220**
Wolf 5	261.3378	−26.1069	73.3246	170.0670	336.9727	55.2653
Wolf 6	57.4357	291.2658	−226.0236	151.2887	268.4628	59.5353

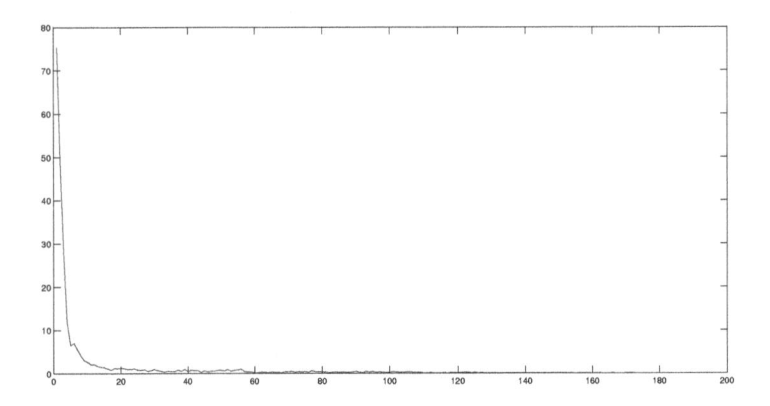

Figure 8.4: Alpha Variation over Iterations

9 Teaching Learning Based Optimization

9.1 INTRODUCTION

Teaching Learning Based Optimization (TLBO) is an optimization technique based on the process of improvement of the performance of a learner in a class. The performance of a learner depends on the quality of the lectures delivered by the faculty, that is, the quality of the teacher. The teacher is assumed to have the collection of knowledge. The performance of the learner also improves with his/her interaction with other learners. Thus, the performance of a learner can be optimized with his/her interactions with the faculty/teacher and the learner.

TLBO was introduced in [127] and [128] by R. Venkata Rao et al. in the years 2011 and 2012, respectively. The method has not been implemented as widely as PSO or GA but presents a different approach to solving the EOA problem. The optimization technique divides the process into two parts – teacher phase, where the influence of the teacher helps the learner to improve, and learner phase, where the interaction amongst the learners help a learner improve his/her knowledge.

Even though the concept of the method is very easy to understand and no mathematically complex calculations are involved, a programmer has to put in a lot of effort to combine the two phases in the method and achieve the output. However, the better part is that the algorithm does not have any constant value that needs to be fixed.

A very good source of information related to TLBO and variations in it, is available at [126]. Some of the drawbacks of TLBO are critically analyzed in [24]. These findings are listed below:

- TLBO cannot be considered as parameter-less algorithm.
- TLBO cannot be implemented very easily.
- TLBO does not require less computational efforts.

TLBO also has some idealistic assumptions, which may practically not exist, like:

- One of the learners becomes the teacher, and then in the next iteration another learner becomes the teacher.
- Teachers are compared only on the basis of marks obtained by the learners. (What if one of the teacher is lenient, but not good?)
- Learners try to learn from other learners, but:

  Practically the number of learners with whom one interacts cannot be fixed or constant.
  Learners who are reserved in nature may not have interactions with anyone at all.

- Learners always improve through interactions.

DOI: 10.1201/b22647-9

• A teacher's knowledge cannot change based on the mean of the class, but this happens in the Teacher Phase calculation as explained in [126].
• A teacher again becomes a learner during the students phase.
• According to [126], two learners having the same objective function value, i.e., performance level, cannot interact.
• A learner is prohibited from interacting with another learner having the same output [126], even though their marks in different subjects can vary.
• The interaction with a teacher or another learner affects all the subject marks of a learner.
• When one learner interacts with another, the other learner should also get influenced, something like mutual coupling in network analysis. In other words, the influence can be negative as well, or a bad learner can also influence a good learner.

The TLBO method is still a promising optimization technique and the following section provides exploration into this method.

So let's enter the class!

9.2 TERMINOLOGY

This section tries to summarize a collection of terms used in TLBO along with its interpretation in the optimization technique, based on the available information:

1. **Learner**: A probable solution to the optimization problem.
2. **Teacher**: The learner solution having best optimization function value at the beginning of Teacher phase calculations.
3. **Subject**: One input factor/variable. A collection of subjects shall define the different aspects of search space.
4. **Class**: A collection of learners/possible solutions. Similar to swarm in PSO.
5. **Marks/Grade/Output**: It is the output or the evaluated objective function value of a possible solution.
6. **Teaching Factor**: It is useful in deciding the value of mean to be changed.
7. **Teacher Phase**: A phase in TLBO in which the learners/solutions are influenced by the teacher/best solution.
8. **Student/Learner Phase**: A phase in TLBO in which the learners/solutions are influenced by other learners/solutions.

9.3 FUNDAMENTAL CONCEPT

In this section, let us understand how TLBO functions and how the different equations and phases are implemented in the method.

TLBO, like any other EOA, considers multiple solutions for finding the optimal solution. In TLBO, the solutions are represented by learners, i.e., like a particle in PSO. A collection of learners makes up a class. The output of the learners is given by the cumulative effect of marks obtained in different subjects. In other words, the objective function value of the learner is the total marks obtained by the learner. Each input factor or variable is considered a subject. TLBO is divided into two phases: the Teacher phase and the Learner/Student phase.

Teacher Phase

In Teacher Phase the learners are influenced by the teacher. Initially, at the start of the process, a number of learners i.e., random solutions, are generated. The collection of these solutions is called a Class. The objective function values of each solution in a class are evaluated. The solution having the best objective function value in the class is designated as a Teacher. In the Teacher phase the learners are influenced by the Teacher. The Teacher tries to improve the knowledge of the learners and get them near to his/her own level of knowledge. The influence is obtained as given in Eq 9.1:

$$Difference\_Mean_{j,k,i} = r_{j,i}(X_{j,kbest,i} - T_F M_{j,i}) \quad (9.1)$$

Eq 9.1 has been referred from [126]. But it was thought that some small changes are required to make the equation more easy to understand. In Eq 9.1,

i $\rightarrow$ iteration number
j $\rightarrow$ one of the subjects i.e., input variable, j = 1, 2, ...m
k $\rightarrow$ one of the learners, k= 1, 2, ...n
$r_{j,i}$ $\rightarrow$ random number for a given subject and iteration. Its value should be between 0 and 1
$X_{j,kbest,i}$ $\rightarrow$ marks of the Teacher i.e., best learner/solution for subject 'j'
$M_{j,i}$ $\rightarrow$ Mean result of learners in subject 'j'
T_F $\rightarrow$ Teaching Factor

The teaching factor can be calculated as given in Eq 9.2

$$T_F = round[1 + rand(0,1)2 - 1] \quad (9.2)$$

In Eq 9.2, the value of T_F is randomly decided and its value can be 1 or 2.

The $Difference\_Mean_{j,k,i}$ is evaluated once for each subject in the Teacher phase of TLBO. This $Difference\_Mean$ is then added to the respective subject marks of all learners in the class, including the teacher as shown in Eq 9.3.

$$X'_{j,k,i} = X_{j,k,i} + Difference\_Mean_{j,k,i} \quad (9.3)$$

The new objective function values for the new solutions '$X'_{j,k,i}$' are evaluated for every learner in the class. If there is an improvement in the objective function value, the new solution $X'_{j,k,i}$ is stored for the respective learner, otherwise the old solution $X_{j,k,i}$ is retained. This ends the Teacher phase of TLBO.

Learner/Student Phase

The student phase of TLBO concentrates on the influence of learners on their peers. For evaluating these effects, all the learners go through one of Eqs 9.4a or 9.4b. If the learner under consideration has a better objective function value than the learner with whom it is interacting, then it goes through Eq 9.4a. Otherwise in a reverse case, where the learner under consideration has an objective function value which is not better than the learner with whom it is interacting, then it goes through Eq 9.4b. The equations have been taken from [126], which presents a modified and more correct version of the equations as compared to the initial publications on TLBO. However, these equations also have been slightly modified for more accurate representation.

$$X''_{j,P,i} = X'_{j,P,i} + r_{j,i}\left(X'_{j,P,i} - X'_{j,Q,i}\right) \quad (9.4a)$$

$$X''_{j,P,i} = X'_{j,P,i} + r_{j,i}\left(X'_{j,Q,i} - X'_{j,P,i}\right) \quad (9.4b)$$

In Eqs 9.4a and 9.4b,

i $\rightarrow$ iteration count
P $\rightarrow$ learner under consideration
Q $\rightarrow$ learner with whom interaction is taking place
$r_i \rightarrow$ random number between 0 and 1 for subject 'j' for 'i^{th}' iteration.
$X''_{j,P,i} \rightarrow$ new value of subject 'j' for 'P' learner in 'i^{th}' iteration
$X'_{j,P,i} \rightarrow$ old value of subject 'j' for 'P' learner in 'i^{th}' iteration after Teacher phase
$X'_{j,Q,i} \rightarrow$ old value of subject 'j' for 'Q' learner in 'i^{th}' iteration after Teacher phase

Before making these evaluations, the following constraints should be satisfied:

- $i \neq j$
- Objective function values of both the learners should not be the same

Eq 9.4a is used when the learner has a better objective function value as compared to the learner from whom it is getting influenced, whereas Eq 9.4b, is used when the objective function value of the learner who is influencing is better than the one that is getting influenced. In these equations, 'P' is the learner getting influenced and 'Q' is the learner who is influencing 'P'.

The $X''_{j,P,i}$ value for all values of 'j' shall be accepted if, all the new subject values of the learner give a better objective function value. Otherwise the old subject values are retained.

After finding the new solutions for every learner, their objective function values are evaluated. The learners are updated by taking up new marks if they are better than the previous ones or retaining the old marks if they are better than the new marks.

This ends, one iteration of TLBO. The iterations are repeated until the stopping criterion is met. Before moving to the next iteration, the best learner in the class will be nominated as the teacher.

9.4 ALGORITHM AND PSEUDOCODE

In TLBO, like any other optimization technique, the input data is first acquired. Then the random solutions, i.e., the learners are generated taking into consideration that the input variables are within limits. If not then the input variables are fixed to the prespecified limits.

The TLBO process is explained through 3 flowcharts i.e., the main process, the teacher phase and the learner phase. Algorithm 9.1 gives the main flowchart of TLBO.

Algorithm 9.1 Teaching Learning Based Optimization

```
START
Input Required Data
for 1 to Population Size do
   Generate Learner/Student/Solution
   if Marks > Max Limit or Marks < Min Limit then
      Marks = Limit
   end if
end for
for 1 to Population Size do
   Obtain Objective Function Value
end for
for 1 to Maximum Number of Iterations do
   Teacher Phase (Algorithm 9.2)
   Learner/Student Phase (Algorithm 9.3)
end for
Display Result
STOP
```

Once the random Learners have been generated, their objective function values are evaluated. In the last loop of Algorithm 9.1, the Teacher phase and Learner phase are executed in each iteration until the stopping criterion has been satisfied.

The Teacher phase of TLBO is represented in Algorithm 9.2.

Algorithm 9.2 Teacher Phase

```
START
Find Teacher i.e., Learner with best objective function value in the class
Find Average of each subject
Obtain Difference_Mean, random numbers and T_F can be generated dynamically
for 1 to Population Size do
   Add Learner and Difference_Mean to get new Learner Marks
   if Marks > Max Limit or Marks < Min Limit then
      Marks = Limit
   end if
end for
for 1 to Population Size do
   Obtain Objective Function Value
end for
for 1 to Population Size do
   if New Learner is better than Old Learner then
      Replace the Old Learner with New Learner
   end if
end for
STOP
```

In the Teacher phase of TLBO, the best Learner within the class is identified, based on the objective function value and is termed the Teacher. In the next part, the average marks of each subject i.e., input variables, is found. After that, the $Difference\_Mean$ (Eq 9.1), random numbers and T_F (Eq 9.2) are generated dynamically. The Learner marks are now updated using Eq 9.3 through a loop to cover all the Learners. The limits are checked for the new Learner marks and their objective function values are found. If the new Learner marks are better than the old marks then, the Learner marks are updated, otherwise the old marks are retained. This ends the Teacher phase of TLBO.

The next phase in the TLBO process is the Learner phase. The process of Learner phase is given in Algorithm 9.3. In the initial part of this phase random numbers are generated for each subject, i.e., input variable. Then a loop is repeated through the class for each learner. In this loop, for each Learner, an influencer is identified, randomly. If the randomly found influencing Learner is the same as the Learner in consideration or the influencing Learner has the same objective function value as the Learner being influenced, then another Learner needs to be identified for the pair. The process of identifying suitable Learner pairs is repeated until the conditions are satisfied.

The new marks for the Learner phase is obtained for each Learner using Eq 9.4a and 9.4b. The limits for the new marks are checked and their objective function values are obtained. The old Learner marks shall be replaced with the new Learner marks if the objective function value for the new marks is better than the old marks, otherwise the old marks are retained. This process end the Learner phase of TLBO.

The Teacher phase and Learner phase are alternately executed in TLBO, until the stopping criterion is satisfied.

9.5 FLOWCHART

The full flowchart for TLBO is presented in Fig 9.1. The flowchart gives broadly the steps involved in the optimization process. Each step can further be expanded into its own flowchart.

Initially, the learners are generated randomly. The subject marks are the values of the input factors/variables. It should be kept in mind that the generation of marks should be within the acceptable limits. The number of learners in the class should be predefined. Once the learners are generated, their objective function values should be evaluated. This forms the basis to start the search process.

The TLBO process evolves around two phases for achieving optimal results. These steps are Teacher phase and Learner/Student phase. In the Teacher phase, the learners are influenced by a single solution which has the best objective function value in the class and is termed Teacher. The Teacher phase is explained in Fig 9.2. The next important stage in TLBO is the Learner/Student phase where the learners are influenced by other learners. These learners who influence others are randomly selected. The Learner/Student phase is explained in Fig 9.3. The Teacher and Learner/Student phase are represented as single blocks in the main flowchart, but later expanded and explained in detail.

Algorithm 9.3 Learner/Student Phase

START
for 1 **to** Number of Subjects **do**
  Generate Random Weights
end for
for 1 **to** Population Size **do**
  Generate Random Pair
  if Index **or** Objective Function Value of Current Learner and Paired Learner are equal **then**
    Repeat the Pairing
  end if
end for
for 1 **to** Population Size **do**
  Obtain New Learner Marks using Eq 9.4a and Eq 9.4b
  if Marks > Max Limit **or** Marks < Min Limit **then**
    Marks = Limit
  end if
end for
for 1 **to** Population Size **do**
  Obtain Objective Function Value
end for
for 1 **to** Population Size **do**
  if New Learner is better than Old Learner **then**
    Replace the Old Learner with New Learner
  end if
end for
STOP

After going through the Teacher phase and the Learner/Student phase, the termination criterion is checked. In [126], which is a very dependable source of information related to TLBO, the termination criterion is always assumed to be "number of iterations". However, similar to PSO, the attainment of equal level of knowledge among all learners of the class can also be assumed to be a stopping criterion, which would correspond to convergence of particles in PSO. Another stopping criterion can be, no improvement in the knowledge level of the learners for a specified number of iterations.

If the stopping criterion is achieved then the results are displayed and the optimization process is stopped. Otherwise the TLBO will again go through, the Teacher phase and Learner/Student phase and the loop continues until stopping criterion is met.

Teacher Phase

The Teacher phase of TLBO is represented through a flowchart in Fig 9.2. The first step in the Teacher phase is to find the Teacher. The Teacher is the learner having the best objective function value in the class. The next step requires us to generate

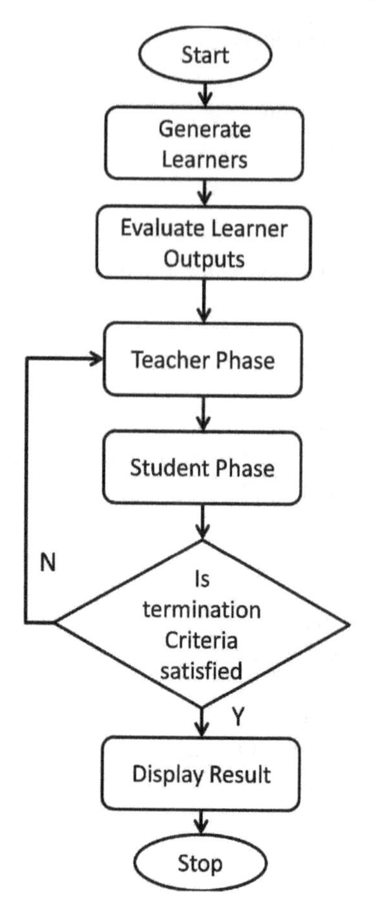

Figure 9.1: TLBO Flowchart

random numbers for each subject, these random numbers should be between 0 and 1. It is also required that the mean of each subject is obtained. The base for the calculations in Teacher phase is reached with this step. Next the $Difference\_Mean$ is to be found for each subject using Eq 9.1. The subject marks of the class are then upgraded through the Eq 9.3, in which the $Difference\_Mean$ is added to the respective subjects of each learner. The objective function values for the newly obtained marks of each learner is then found. In the last step, a comparison is done between the objective function values of each learner before the Teacher phase and the newly obtained values. The solution having better objective function value is retained for each learner. This ends the Teacher phase of TLBO.

Learner Phase

The Learner/Student phase of TLBO has got lesser steps, but it appears to be more complicated in coding as compared to Teacher phase. The flowchart for Learner/Student phase is given in Fig 9.3. This phase of TLBO starts with the generation

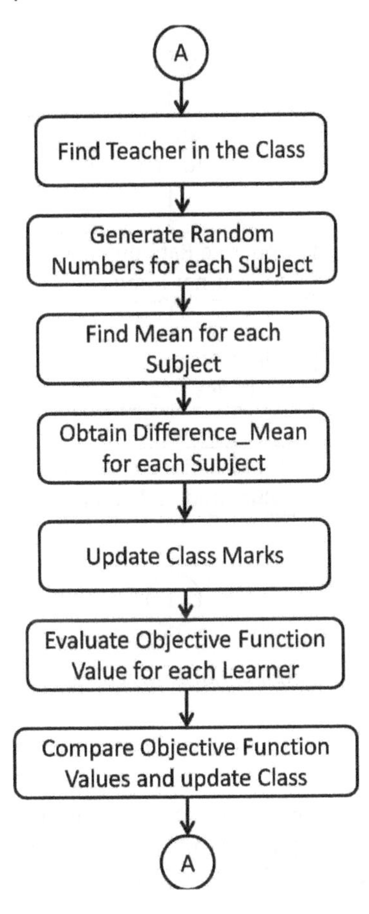

Figure 9.2: Teacher Phase

of random numbers. The random numbers generated should be equal to the number of subjects. It should also be noted that the random numbers cannot be generated dynamically, like in other EOA, as they remain constant for each subject. Next part of the Learner/Student phase requires generation of pairs for interaction. The influencer being paired with the learner should have a different index as well as objective function value. Once the pairs are generated, the interaction takes place between the learner and the influencer, using Eqs 9.4a and 9.4b. This step creates a new set of marks for the class. The objective function values for this new set of marks are obtained and compared with the set of marks of learners after the Teacher phase. If the objective function value of a learner improves for the new set of marks then these marks are adopted, otherwise it retains the previous marks. During the whole process, it should be kept in mind that whenever a new set of marks are being generated, the range of the marks should be checked such that they fall in the prescribed limits. This ends the Learner/Student phase.

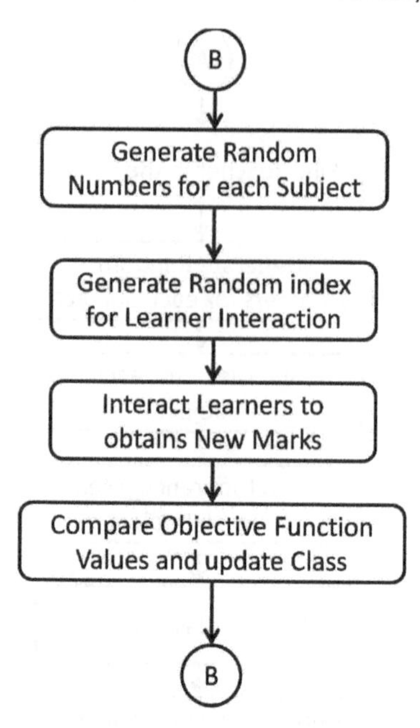

Figure 9.3: Learner Phase

9.6 EXAMPLE

This section shall solve the Griewank Optimization Function given by Eq 2.5. The solution of the equation is explained in Eq 2.9. The teacher phase and the learner phase shall be explained in detail with step-by-step solution to the problem for two iterations.

The number of input variables are assumed to be 5, i.e., the number of subjects are considered to be 5. The number of learners/solutions in a class is taken as 6. The range of each subject marks is from −600 to 600. Let the learners/solutions be represented by the values in Table 9.1

Table 9.1
Initial Class of TLBO

	Subject 1	Subject 2	Subject 3	Subject 4	Subject 5
Learner 1	141.8811	−104.656	−397.606	211.7688	561.0059
Learner 2	−16.0475	−277.517	74.94419	247.9742	391.221
Learner 3	503.4618	202.9188	51.23069	356.5671	321.7123
Learner 4	−80.9499	−329.007	−180.324	77.46379	397.8965
Learner 5	471.3352	−152.64	86.91336	386.1857	−460.428
Learner 6	−556.071	286.2077	118.7875	249.8144	106.6543

The learners are evaluated for their objective function values using Eqs 2.5 and 2.9. These objective function values are given in Table 9.2

Table 9.2
Objective Function Values for the Learners

	Objective Function Value	Remark
Learner 1	138.1064	
Learner 2	75.3538	Teacher
Learner 3	132.9249	
Learner 4	78.7628	
Learner 5	154.5424	
Learner 6	120.6237	

It can be observed from Table 9.2, that the objective function value of the Learner 2 is minimum and thus the best in the class. Thus Learner 2 will behave like a Teacher in the Teachers phase of the first iteration.

Teacher Phase

In Teacher phase, there is a need to find $Difference\_Mean$, which further requires the mean of all subjects. So the first step is to find the means of all the subjects, please refer Table 9.1. The mean of the subject marks are given in Table 9.3.

Table 9.3
Mean Marks of the Class

	Subject 1	Subject 2	Subject 3	Subject 4	Subject 5
Mean	77.2683	−62.4488	−41.009	254.9623	219.6771

To find the $Difference\_Mean$, according to Eq 9.1, we are also required to generate random numbers. Let the generated random numbers be as shown in Table 9.4

Table 9.4
Random Numbers for Different Subjects in Teacher Phase

	Subject 1	Subject 2	Subject 3	Subject 4	Subject 5
Random Numbers	0.58	0.44	0.2	0.8	0.91

By taking more number of input variables, we may be able to analyze the effect of random numbers on the process. The most important calculation in the Teacher

Phase is to find the $Difference\_Mean$. According to Eq 9.1, the $Difference\_Mean$ for the first subject can be obtained as:

$Difference\_Mean = 0.58 * (-16.0475 - (1 * 77.2683)) = -54.1232$

We will be assuming the value of T_F to be 1, throughout the iteration. It is up to the reader to implement various values for T_F and random numbers, to observe their contribution in the optimization process. In the above calculation,

0.58 $\rightarrow$ random number for subject 1, refer Table 9.4
−16.0475 $\rightarrow$ subject 1 marks of the Teacher, refer Table 9.1
1 $\rightarrow$ Teaching factor, value assumed to be 1.
77.2683 $\rightarrow$ Mean value of subject 1, refer Table 9.3

Similarly, the $Difference\_Mean$ values for all the subjects are calculated and presented in Table 9.5.

Table 9.5
Difference_Mean of the Class

	Subject 1	Subject 2	Subject 3	Subject 4	Subject 5
Difference_Mean of the Class	−54.1232	−94.6298	23.19065	−5.59049	156.105

In the next step, we need to add the values, obtained in Table 9.5, to each value of the respective columns of Table 9.1. Thus, −54.1213 will be added to all the values of the first column of Table 9.1, −94.6298 shall be added to all the values of second column and so on.

The updated values of the subject marks for all learners is given in Table 9.6.

Table 9.6
Updated Class of TLBO

	Subject 1	Subject 2	Subject 3	Subject 4	Subject 5
Learner 1	87.758	−199.286	−374.416	206.1783	**600**
Learner 2	−70.1707	−372.147	98.13484	242.3837	547.3259
Learner 3	449.3386	108.2889	74.42134	350.9766	477.8172
Learner 4	−135.073	−423.637	−157.133	71.8733	554.0015
Learner 5	417.212	−247.27	110.104	380.5952	−304.323
Learner 6	**−600**	191.5779	141.9781	244.2239	262.7592

For Learner 1 and Learner 6, the subject 5 and subject 1 values, are going out of range and are therefore restricted to the maximum (600) and minimum range (−600) values, respectively. These values are marked as bold in the table.

The objective function values for these updated marks have been evaluated in Table 9.7.

Table 9.7
Updated Objective Function Values for the Learners

	Objective Function Value
Learner 1	148.3621
Learner 2	128.9232
Learner 3	143.8495
Learner 4	134.6938
Learner 5	122.2336
Learner 6	137.1455

A comparison between the updated objective function values after the Teacher phase and that before the Teacher phase are presented in Table 9.8. The objective function values which are better are selected and retained. It can be observed from the comparison that only one learner, i.e., the 5th, could be updated in the whole class after the end of Teacher phase. The objective function value for the 5th learner improves from 154.5424 to 122.2336. The table only shows the learner marks that would be chosen for the Learner phase. It is suggested to the readers to experiment and analyze with the random values used during the Teacher phase and observe the number of learners getting improved in the respective phase.

The lower part of Table 9.8, gives the marks and objective function values of the learners to be taken to the Learner/Student phase.

Learner/Student Phase

In the Learner/ Student phase, the learners are influenced by other learners. There is no Teacher in this phase of calculation. The learner is usually influenced by another learner, which is randomly selected. The level of influence is decided by a random number having value between 0 and 1. The equations used in this phase are 9.4a and 9.4b. Both the equations are similar and easy to implement.

The random numbers are generated based on the number of subjects. Let these random numbers be as given in Table 9.9.

The learners are supposed to interact with other learners during the learner phase. The two conditions that this interaction process should satisfy are

(i) the learner cannot interact with himself/herself

(ii) the learner cannot interact with another learner with the same objective function value [126]

In the class that we obtain at the end of the Teacher phase it can be observed that no two learners have the same objective function value. Thus the second condition shall be satisfied anyhow. Let Table 9.10 give the combination of learners and the respective influencers with whom they interact.

From Table 9.10, it can be observed that none of the learner–influencer pairs have the same values. Thus condition 1 is also satisfied. It should also be seen that learners 4 and 5 are influencing two learners each. The pairs have been randomly generated.

Table 9.8
Comparison before and after Teacher Phase of TLBO

	Learner Values after Teacher phase calculation						Learner Values before Teacher phase calculation					
	Subject 1	Subject 2	Subject 3	Subject 4	Subject 5	Objective Func Value	Objective Func Value	Subject 1	Subject 2	Subject 3	Subject 4	Subject 5
Learner 1						148.3621	**138.1064**	141.8811	−104.656	−397.606	211.7688	561.0059
Learner 2						128.9232	**75.35377**	−16.0475	−277.517	74.94419	247.9742	391.221
Learner 3						143.8495	**132.9249**	503.4618	202.9188	51.23069	356.5671	321.7123
Learner 4						134.6938	**78.76276**	−80.9499	−329.007	−180.324	77.46379	397.8965
Learner 5	−54.1232	−94.6298	23.19065	−5.59049	156.105	**122.2336**	154.5424					
Learner 6						137.1455	**120.6237**	−556.071	286.2077	118.7875	249.8144	106.6543

Learner 1	141.8811	−104.656	−397.606	211.7688	561.0059	**138.1064**
Learner 2	−16.0475	−277.517	74.94419	247.9742	391.221	**75.35377**
Learner 3	503.4618	202.9188	51.23069	356.5671	321.7123	**132.9249**
Learner 4	−80.9499	−329.007	−180.324	77.46379	397.8965	**78.76276**
Learner 5	−54.1232	−94.6298	23.19065	−5.59049	156.105	**122.2336**
Learner 6	−556.071	286.2077	118.7875	249.8144	106.6543	**120.6237**

Table 9.9
Random Numbers for Different Subjects in Learner Phase

	Subject 1	Subject 2	Subject 3	Subject 4	Subject 5
Random Numbers	0.81	0.92	0.2	0.05	0.4

Table 9.10
Learner Influencer Pair for Learner Phase in Iteration 1

Learner	Influencer
1	4
2	5
3	6
4	5
5	4
6	2

A question comes up here, should the chance for each learner to be an influencer be dependent on its objective function value? Can the different selection process of GA be applied for selection of learners to be influencers? Will it improve the process? Because in that case the learners who are having better marks/objective function value will have a better chance to influence others and improve the class mean as well. It would be therefore that learner 2 gets more chance to influence others. This is for the readers to evaluate.

Let us now start with the Learner/Student phase by calculating the new marks for learner 1. The learner 1 is influenced by learner 4, refer Table 9.10. From Eqs 9.4a and 9.4b, the new marks for learner 1 can be evaluated as:

Learner 1, Subject 1, Marks $= 141.8811 + 0.81 * (-80.9499 - 141.8811) = -38.612$

Learner 1, Subject 2, Marks $= -104.6557 + 0.92 * (-329.007 - (-104.6557)) = -311.059$

In the above calculations,

141.8811 and −104.6557 → Marks of learner 1 for subjects 1 and 2, respectively. Refer to Table 9.8
−80.9499 and −329.007 → Marks of learner 4 (influencer) for subjects 1 and 2, respectively. Refer to Table 9.8
0.81 and 0.92 → Random weights for subject 1 and 2, respectively. Refer to Table 9.9.

Since the objective function value of influencer, i.e., learner 4 (78.76276) is better than learner 1 (138.1064), who is getting influenced, Eq 9.4b is used. Similar calculation shall be done for the remaining subjects of learner 1. Now let's try for learner 2, who is influenced from learner 5, refer Table 9.10. For obtaining the marks of learner 2, Eq 9.4a would be used, since the objective function value of learner 2 (75.35377) is better than the influencer, i.e., learner 5 (122.2336).

$$\text{Learner 2, Subject 1, Marks} = -16.0475 + 0.81 * (-16.0475 - (-54.1232)) = 14.7938$$

$$\text{Learner 2, Subject 2, Marks} = -277.517 + 0.92 * (-277.517 - (-94.6298)) = -445.7732$$

Similar approaches shall be taken to obtain the remaining marks of learner 2 and the remaining learners. It should be kept in mind that after every calculation, the limits of the variables should be checked.

Table 9.11 gives the marks obtained for all the learners after the Learner/Student phase.

Table 9.11
Learner/Student Phase Marks for the Class for TLBO

	Subject 1	Subject 2	Subject 3	Subject 4	Subject 5
Learner 1	−38.612	−311.059	−354.15	205.0536	495.7621
Learner 2	14.79383	−445.773	85.2949	260.6525	485.2674
Learner 3	−354.76	279.5446	64.74205	351.2295	235.6891
Learner 4	−102.68	−544.634	−221.027	81.6165	494.6131
Learner 5	−75.8528	−310.257	−17.5122	−1.43778	252.8216
Learner 6	−118.652	−232.419	110.0188	249.7224	220.4809

The objective function values for the above class is found and given in Table 9.12.

Table 9.12
Updated Objective Function Values of Class during Learner/Student Phase

	Objective Function Value
Learner 1	128.9297
Learner 2	128.4155
Learner 3	97.64039
Learner 4	152.8164
Learner 5	43.01213
Learner 6	48.87149

The objective function values of the class before Learner/Student phase and after obtaining the new marks are evaluated. The learners with better objective function values are chosen to be a part of the class. Table 9.13 gives a comparison between these two stages while the lower part of the table gives the final learner population.

Table 9.13
Comparison before and after Learner/Student Phase of TLBO

	Learner Values after Learner phase calculation							Learner Values before Learner phase calculation				
	Subject 1	Subject 2	Subject 3	Subject 4	Subject 5	Objective Func Value	Objective Func Value	Subject 1	Subject 2	Subject 3	Subject 4	Subject 5
Learner 1	−38.612	−311.059	−354.15	205.0536	495.7621	**128.9297**	138.1064					
Learner 2						128.4155	**75.35377**	−16.0475	−277.517	74.94419	247.9742	391.221
Learner 3	−354.76	279.5446	64.74205	351.2295	235.6891	**97.64039**	132.9249					
Learner 4						152.8164	**78.76276**	−80.9499	−329.007	−180.324	77.46379	397.8965
Learner 5	−75.8528	−310.257	−17.5122	−1.43778	252.8216	**43.01213**	122.2336					
Learner 6	−118.652	−232.419	110.0188	249.7224	220.4809	**48.87149**	120.6237					

Learner 1	−38.612	−311.059	−354.15	205.0536	495.7621	**128.9297**
Learner 2	−16.0475	−277.517	74.94419	247.9742	391.221	**75.35377**
Learner 3	−354.76	279.5446	64.74205	351.2295	235.6891	**97.64039**
Learner 4	−80.9499	−329.007	−180.324	77.46379	397.8965	**78.76276**
Learner 5	−75.8528	−310.257	−17.5122	−1.43778	252.8216	**43.01213**
Learner 6	−118.652	−232.419	110.0188	249.7224	220.4809	**48.87149**

From Table 9.13 it can be observed that learner 5 has achieved the minimal objective function value of 43.0121. The TLBO is further solved for one more iteration, however, the calculations are not explained. It is for the reader to confirm the calculations and get more familiar with the optimization method. Table 9.14 shows the calculations for Teacher phase of iteration 2. Learners 1, 2, 3 and 6 improve their objective function values in this iteration.

Table 9.14
Teacher Phase of Second Iteration for TLBO

Learner	Subject 1	Subject 2	Subject 3	Subject 4	Subject 5	Obj. Fun. Value	Remarks
1	−38.612	−311.059	−354.15	205.0536	495.7621	128.9297	
2	−16.0475	−277.517	74.94419	247.9742	391.221	75.35377	
3	−354.76	279.5446	64.74205	351.2295	235.6891	97.64039	
4	−80.9499	−329.007	−180.324	77.46379	397.8965	78.76276	
5	−75.8528	−310.257	−17.5122	−1.43778	252.8216	43.01213	Teacher
6	−118.652	−232.419	110.0188	249.7224	220.4809	48.87149	
	−114.146	−196.786	−50.3801	188.3343	332.3119		Mean
	0.94	0.54	0.62	0.88	0.12		Random Numbers
	35.99521	−61.2745	20.37808	−166.999	−9.53883		Difference Mean
1	−2.61682	−372.333	−333.772	38.05416	486.2233	123.2386	Marks obtained in Teacher phase
2	19.94773	−338.791	95.32228	80.97482	381.6821	70.13385	
3	−318.764	218.2701	85.12014	184.23	226.1502	61.41441	
4	−44.9547	−390.281	−159.946	−89.5356	388.3577	85.6183	
5	−39.8576	−371.531	2.865848	−168.437	243.2827	57.79164	
6	−82.6567	−293.693	130.3969	82.72299	210.9421	41.81369	
1	−2.61682	−372.333	−333.772	38.05416	486.2233	123.2386	Final Marks after Teacher phase
2	19.94773	−338.791	95.32228	80.97482	381.6821	70.13385	
3	−318.764	218.2701	85.12014	184.23	226.1502	61.41441	
4	−80.9499	−329.007	−180.324	77.46379	397.8965	78.76276	
5	−75.8528	−310.257	−17.5122	−1.43778	252.8216	43.01213	
6	−82.6567	−293.693	130.3969	82.72299	210.9421	41.81369	

From Table 9.14, we can observe that the optimal objective function value at the beginning of the Teacher phase was 43.01213, which has now been reduced to 41.81369. The end of Teacher phase starts the Learner/Student phase. The Learner/Student phase calculations are given in Table 9.15. The initial value of marks of the class is the same as where it was left by the Teacher phase.

The second set of marks in Table 9.15 are obtained during the Learner/Student phase. From the table, it can be observed that the learners 1, 4 and 5 have improved

Table 9.15
Learner Phase of Second Iteration for TLBO

Learner	Subject 1	Subject 2	Subject 3	Subject 4	Subject 5	Obj. Fun. Value	Remarks
1	−2.61682	−372.333	−333.772	38.05416	486.2233	123.2386	Start Learner phase
2	19.94773	−338.791	95.32228	80.97482	381.6821	70.13385	
3	−318.764	218.2701	85.12014	184.23	226.1502	61.41441	
4	−80.9499	−329.007	−180.324	77.46379	397.8965	78.76276	
5	−75.8528	−310.257	−17.5122	−1.43778	252.8216	43.01213	
6	−82.6567	−293.693	130.3969	82.72299	210.9421	41.81369	
	0.02	0.44	0.63	0	0.45		Random Weights
1	−8.93977	−112.468	−69.8698	38.05416	369.1904	39.76749	Pair 1 - 3
2	20.39902	−324.033	365.6515	80.97482	334.6386	90.42016	Pair 2 - 1
3	−325.087	478.1357	349.022	184.23	109.1174	126.4886	Pair 3 - 1
4	−78.932	−333.312	−6.66676	77.46379	390.6	70.09921	Pair 4 - 2
5	−75.9889	−302.969	75.67053	−1.43778	233.9758	40.79002	Pair 5 - 6
6	−82.6908	−278.155	326.1509	82.72299	126.8126	54.22937	Pair 6 - 4
1	−8.93977	−112.468	−69.8698	38.05416	369.1904	39.76749	Final Marks after Learner phase
2	19.94773	−338.791	95.32228	80.97482	381.6821	70.13385	
3	−318.764	218.2701	85.12014	184.23	226.1502	61.41441	
4	−78.932	−333.312	−6.66676	77.46379	390.6	70.09921	
5	−75.9889	−302.969	75.67053	−1.43778	233.9758	40.79002	
6	−82.6567	−293.693	130.3969	82.72299	210.9421	41.81369	

their marks and the optimal objective function value has further reduced to 39.76749. The optimization method can be repeated for a number of iteration or until all learners have the same marks, to obtain the optimal solution.

The program of TLBO was executed for 200 times. The variation in the position of the Teacher is plotted for the execution in Fig 9.4. The objective function value at the end of 200 iterations was 0.1612. This method of showing the performance of an EOA is presented in many number of research papers and would be helpful for young researchers. The x-axis gives the number of iterations and the y-axis represents the objective function value.

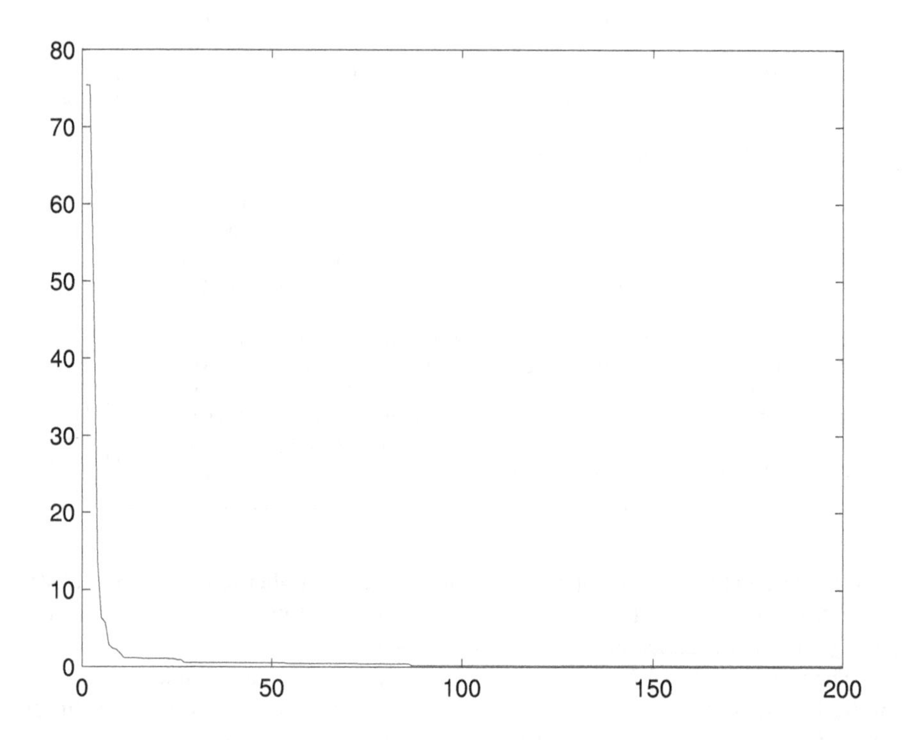

Figure 9.4: Teacher Variation over Iterations

10 Introduction to Other Optimization Techniques

10.1 INTRODUCTION

The chapters before have presented some EOAs in detail. On searching the Internet for optimization techniques, a long list of EOAs can be generated. It was not a possibility to collect all the EOAs, while dedicating one chapter each to a single EOA. Thus, a last chapter has been introduced in this book to briefly present the basic concept, working and formulae related to 6 more EOAs.

This chapter includes optimization techniques as: the Bacteria Foraging Algorithm (2002) based on bacteria foraging and reproduction method, Whale Optimization (2016) based on hunting strategy of humpback whales, the Bat Algorithm (2010) based on hunting strategies of microbats using echolocation, the Firefly Algorithm (2010) based on survival strategies of fireflies, the Gravitational Search Algorithm (2009) based on Newtonian laws and Reducing Variable Trend Search (2014) based on common thought process of multiple individuals. We have taken into account that equations related to each method, pros and cons, are presented in the following sections.

10.2 BACTERIA FORAGING ALGORITHM

Foraging means searching for food. This optimization technique focuses its optimal search process based on the foraging methods employed by bacteria. While foraging for food, an animal has to first locate the food, then pursue / handle it and then eat it. In the food search, an animal optimizes the energy by time ratio, with the contraints of its own abilities (sensing) and external factors (density of prey).

The bacteria foraging method is based on *E. coli* bacteria [111]. The specifications of an *E. coli* bacteria are 1 μm diameter, 2 μm length and 1 pico gm weight. It needs a temperature of around 37°C and has to replicate every 20 mins. It elongates in length and then divides into two 'daughters'. *E. coli* tries to avoid acidic or alkaline regions inside a human where it resides. The foraging behavior of *E. coli* is termed saltatory, which includes taking pauses while moving towards food source.

Chematoxis is a method employed by the bacteria to reach an area of better nutrient concentration. 'θ' gives the position of the bacteria and '$J(\theta)$' represents the objective function value of the bacteria. $J(\theta) < 0$ represents a nutrient rich area whereas $J(\theta) = 0$ represents neutral environment and $J(\theta) > 0$ is for the noxious one. One iteration in bacteria foraging shall include chemotaxis, reproduction and dispersal, which are represented through 'j', 'k' and 'l', respectively. The total number of bacteria used for optimization is given by 'S' and 'N_c' gives the lifetime of the bacteria

DOI: 10.1201/b22647-10

which will be represented through the number of chemotactic steps. 'C(i)' gives the step size for the ith bacteria.

The movement of the bacteria is given by a tumble, as in Eq 10.1.

$$\theta^i(j+1,k,l) = \theta^i(j,k,l) + C(i)\phi(j) \tag{10.1}$$

where '$\phi(j)$' gives the direction of the tumble and 'C(i)' is the size of the step taken. If the objective function value at the new position is better than the previous one, than another step in the same direction is taken. The process repeats until the objective function value keeps improving. However, the maximum number of steps that can be taken is represented by 'N_s'.

Eq 10.1 represents a raw calculation for the bacteria movement. The attractants and repellants play an important role in the movement of the bacteria. The modified equation for bacterial movement using such attractants and repellants is given in Eq 10.2.

$$\begin{aligned} J_{cc}(\theta, P(j,k,l)) &= \sum_{t=1}^{S} J_{cc}^i\left(\theta, \theta^i(j,k,l)\right) \\ &= \sum_{t=1}^{S}\left[-d_{\text{attract}} \exp\left(-w_{\text{attract}} \sum_{m=1}^{p} \left(\theta_m - \theta_m^i\right)^2\right)\right] \\ &\quad + \sum_{t=1}^{S}\left[h_{\text{repellant}} \exp\left(-w_{\text{repellant}} \sum_{m=1}^{p} \left(\theta_m - \theta_m^i\right)^2\right)\right] \end{aligned} \tag{10.2}$$

where 'cc' represents cell-to-cell and 'P' represents position of each bacteria in the population. Constants 'd', 'w' and 'h' represent the depth, width and height, respectively of the attractant or a repellant. Also 'θ_m^i' is the mth input variable of the ith bacteria.

After moving through chemotactic steps (N_c), the bacteria now move towards reproduction steps given by 'N_{re}'. It is assumed that $S_r = \frac{S}{2}$, are the bacteria which has had enough nutrients to survive and reproduce without mutation. These S_r bacteria are the ones having better objective function value in the population. After dividing into two, these newly produced bacteria will be placed in the same location as its parent. The unhealthy bacteria did not get enough nutrient and shall die.

Next in the process are the elimination dispersal event steps given by 'N_{ed}'. Every bacterium within the population shall encounter the elimination dispersal with a probability of 'p_{ed}'.

In this process of bacteria foraging optimization technique, $N_c > N_{re} > N_{ed}$. There are multiple improvements or research opening within the optimization technique and have been distinctly pointed out by the authors in [111].

10.3 WHALE OPTIMIZATION

Whale Optimization Algorithm (WOA) imitates the hunting strategy of humpback whales better known as bubble net hunting strategy [98]. A whale may grow up to

30m and may have a weight of 180t. The whales are intelligent creatures and possess spindle cells, used for judgement. They even have a dialect of their own.

In the bubble net hunting strategy, the whales come from under their prey and start emitting bubbles. They move upward in a spiral manner or a '9' shaped path. This creates a wall of bubbles around the prey. The bubbling starts from around 12 m depth from the prey. Fig 10.1 shows the movement of the humpback whale.

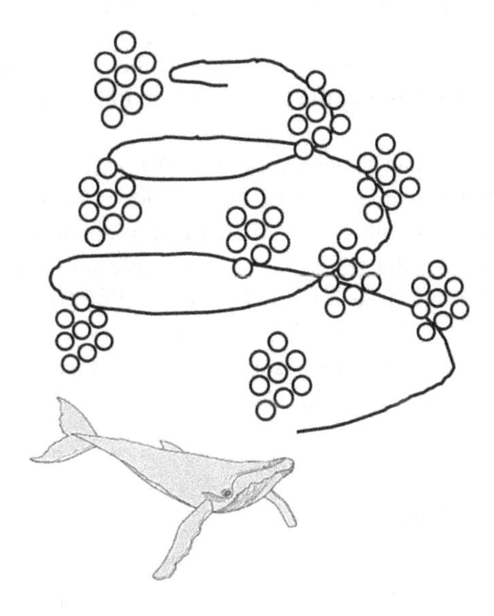

Figure 10.1: Bubble Net Hunting Strategy

The hunting process includes three phases: encircling the prey, bubble netting and exploration. The bubble net is further divided into shrinking encircling and spiral updating position.

During the search process, the best solution is assumed as the prey or the optimal solution. In the encircling part of the search process, Eqs 10.3 and 10.4 are applied:

$$\vec{D} = \left|\vec{C} \cdot \vec{X}^*(t) - \vec{X}(t)\right| \tag{10.3}$$

$$\vec{X}(t+1) = \overrightarrow{X^*}(t) - \vec{A} \cdot \vec{D} \tag{10.4}$$

where '$\vec{X}$' represents the whale/search agent, 'X^*' represents the best solution, 't' represents the iteration number and '$\vec{A}$' and '$\vec{C}$' are coefficient vectors given by Eqs 10.5 and 10.6.

$$\vec{A} = 2\vec{a}.\vec{r} - \vec{a} \tag{10.5}$$

$$\vec{C} = 2.\vec{r} \tag{10.6}$$

where '$\vec{a}$' is reduced from 2 to 0, over the iterations and '$\vec{r}$' is a random vector $\in$ [0,1].

In the first part of bubble net attack, where the shrinking takes place, to reduce the value '$\vec{A}$', the range of '$\vec{a}$' is reduced. '$\vec{A}$' has the range of [-$\vec{a}$, $\vec{a}$].

In the second part of bubble net attack, a spiral equation is created from the current position to the best position. The equation used for the same is given in Eq 10.7.

$$\vec{X}(t+1) = \overrightarrow{D'} \cdot e^{bl} \cdot \cos(2\pi l) + \overrightarrow{X^*}(t) \tag{10.7}$$

where $\overrightarrow{D'} = |\overrightarrow{X^*}(t) - \vec{X}(t)|$ indicating the distance between the ith whale and the prey, 'b' is a constant defining the logarithmic spirals shape and 'l' is a random number $\in [-1,1]$.

The humpback whales can either choose Eq 10.4 or Eq 10.7 for movement. The equation chosen for the movement is based on a random number p $\in$ [0,1]. For p < 0.5, Eq 10.4 is implemented, otherwise Eq 10.7 is implemented.

In the exploration part, the whale moves randomly for searching the prey. The equations used for this random search are the same as Eqs 10.3 and 10.4, except that '$\vec{X}^*(t)$' is replaced by '$\vec{X}_{rand}$' in both the equations.

Some of the clear cons to the optimization technique include the use of too many equations and variables. Also, the method is quite similar to the Grey Wolf Optimizer technique.

10.4 BAT ALGORITHM

Bats are mammals, but they can fly. They have the ability to fly as their forelimbs are converted into wings. Bats are classified broadly into megabats and microbats. Megabats are usually large and eat fruits, whereas microbats hunt for their food. The microbats make use of echolocation technology to find their path and forage for food. The bats vary in weight from a couple of grams to some kilograms while their wingspan range from a couple of centimeters to a couple of meters.

The echolocating bats create special sounds like Sound Navigation and Ranging (SONAR). These sounds are produced by the bats through their voice boxes, however, some use tongues to generate the soundwaves. The emission of soundwaves may occur through the mouth or nostrils. The soundwaves are directed and do not spread out. These soundwaves are incidental on the objects and are reflected back towards the bat. The bat is able to recieve the reflected soundwaves through its ears. The reflected soundwaves allow the bat to accurately see through its ears. It is said that the bats can even identify a hair through the view generated using echolocation. The echolocation helps the bats to hunt in the night. It can easily identify the prey through this method in the absence of light as well.

The echolocation calls have a frequency range from 20 kHz to 200 kHz, which lies in the ultra sonic range and humans are not able to hear these sounds. The duration of these sounds lie in milliseconds (5–20). A bat sends around 10–20 calls per second which may increase to 200 calls per second when the bat is about to catch its prey. The calls may have constant or varying frequency. The low-frequency calls call travel farther as compared to high-frequency calls, however, the high-frequency calls give more details about the environment. The loudness of echolocation calls can vary from 50 dB to 120 dB.

After getting the technical details of echolocating bats, let's understand how this foraging method can be converted into an optimization technique. Figure 10.2 shows the foraging behavior of the bats. The bat and the prey are in the same environment (A), then the bat directs the echolocation call in the direction of the prey (B) and the echolocation calls reflected by the prey are received by the bats (C). The bats then move in the direction of the prey while continuously monitoring the movement of the prey through the calls.

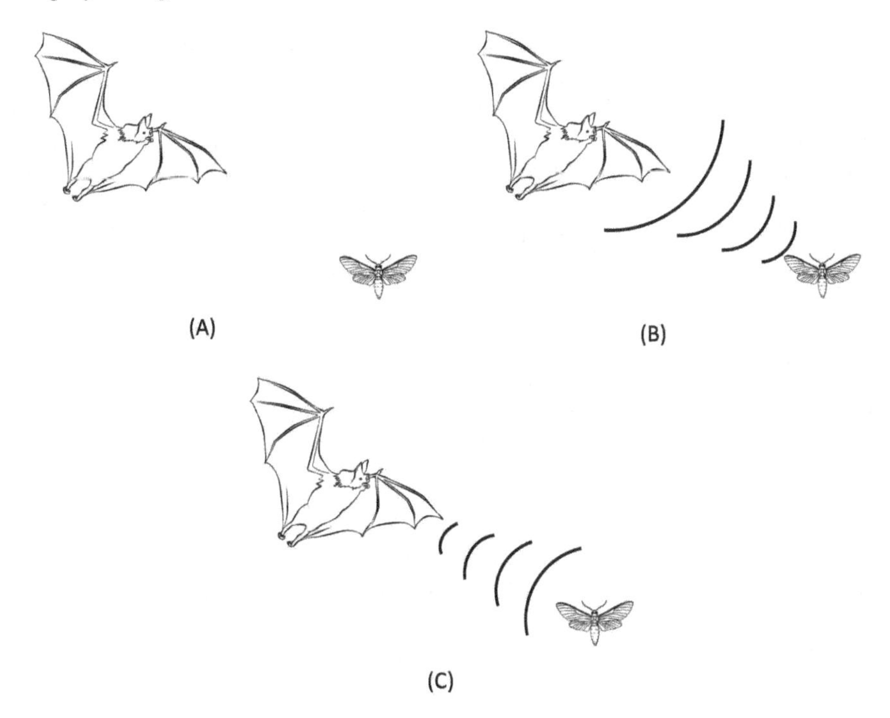

Figure 10.2: Bats Foraging Methodology

The bat-inspired optimization algorithm is introduced in [177]. The bats are virtually assumed to be present. The movement of the bats are defined by Eqs 10.8, 10.9 and 10.10.

$$f_i = f_{\min} + (f_{\max} - f_{\min})\,\beta \tag{10.8}$$

$$\mathbf{v}_i^t = \mathbf{v}_i^{t-1} + \left(\mathbf{x}_i^t - \mathbf{x}_*\right) f_i \tag{10.9}$$

$$\mathbf{x}_i^t = \mathbf{x}_i^{t-1} + \mathbf{v}_i^t \tag{10.10}$$

The frequency 'f_i' is first obtained for the bats using Eq 10.8. The frequency should lie between 'f_{min}' and 'f_{max}'. A random vector 'β' having value in between 0 and 1 is used in this equation. The frequency calculated in Eq 10.8 is then utilized in Eq 10.9 for the evaluation of velocity 'v_i' for the bat. The velocity is the summation of previous velocity and product of frequency and difference between the bats current position 'x_i' and global best location 'x_*'. Eq 10.10 gives the final equation for obtaining new positions for the bat and is the summation of its previous position and the velocity, as in PSO. 't' gives the iteration number.

There are a couple of points to ponder here: (i) The global best position is nowhere to be found in the bat foraging chronology, then why is it implemented in Eq 10.9 and (ii) the Eq 10.9 should have the iteration count as 't-1' for 'x_i'.

There are some more steps involved in the completion of the Bat Algorithm. According to the algorithm presented in the paper, after finding the new position for the bats and updating velocities along with solutions, the pulse emission 'r_i' is matched against a random number. The value of 'r_i' should lie between 0 and 1. If the pulse emission is smaller than the random number, a local solution is generated around the selected best solution. Then another solution is generated by flying randomly. Next a random number is generated, if it is less than the loudness 'A_i' and we obtain a better objective function value for the new positions, then the new solutions are accepted. Along with the acceptance of the new positions, 'r_i' is increased and 'A_i' is reduced.

The equation for generating a new solution for each bat through a random walk is given in Eq 10.11.

$$x_{new} = x_{old} + \varepsilon A^t \tag{10.11}$$

where 'ε' is a random number between -1 and 1 and 'A^t' gives the average loudness of all the bats in that iteration.

The reduction in A_i and increase in r_i is given in Eq 10.12

$$A_i^{t+1} = \alpha A_i^t, \quad r_i^{t+1} = r_i^0[1 - exp(-\gamma t)] \tag{10.12}$$

In Eq 10.12, 'α' and 'γ' are constants, with $0 < \alpha < 1$ and $\gamma > 0$. These values of 'α' and 'γ' are responsible for 'A_i^t' moving towards 0 and 'r_i^t' moving towards 'r_i^0' at $t = \infty$.

The negatives found in the base paper are (i) with the generation of new bats, does the population keeps increasing? and (ii) there are too many constants whose values are not specifically mentioned. The Bat Algorithm has been around for quite some time now and thus it is expected that it must have been through many updates.

10.5 FIREFLY ALGORITHM

The firefly is an insect that uses light signals to communicate. They use the light signals to find mates, attract prey and repel the predators. The light is produced in rhythmic flashes with some patterns being unique to specific species. There are approximately 2000 different species of fireflies. An optimization algorithm was developed around the fireflies in [176].

The basic fundamentals of communication in fireflies are that the light intensity decreases as the distance from the firefly increases. The reduction in light intensity is inversely proportional to the square of the distance of concern.

During the conversion of this firefly communication into an optimization technique, the light intensity can be correlated with the objective function value.

The Firefly Algorithm (FA), has three rules: (1) fireflies are not attracted according to gender, (2) brightness of the flashes are directly proportional to the attractiveness of the firefly and (3) brightness is dependent on the characteristics of objective function being solved.

The movement of the fireflies mainly depends on: light intensity variations and attractiveness. The intensity of light is related to the objective function value, however, attractiveness given by 'β' depends on the relativeness of the brightness between the firefly that emits the flashes and the firefly that sees those flashes. Attractiveness will vary according to the distance between the two fireflies, which makes it as a decreasing function.

The movement of the fireflies is characterized by the Eq 10.13

$$x_i = x_i + \beta_0 e^{-\gamma r_{ij}^2}(x_j - x_i) + \alpha \in_i \tag{10.13}$$

Equation 10.13, shows how the firefly 'i' is attracted to another more brighter firefly 'j'. In the above equation, the second term represents attraction and the third term is for randomness. In the equation, 'β' for implementation purpose, is taken as 1. If 'β' is taken as 0, then the movement of the firefly becomes a random walk. Also, 'γ' is the light absorption coefficient which should ideally be in between 0 and infinity; however, the proposed value is between 0.1 and 10. 'r_{ij}' is a simple cartesian distance between fireflies 'i' and 'j'. 'α' is the randomization parameter and should be between 0 and 1. '$\in_i$' is a vector of random numbers, having uniform distribution. It can be replaced by (rand-1/2) where rand generates random number between 0 and 1, uniformly.

It can be observed that the fireflies are only attracted to the brighter fireflies in its neighborhood.

The FA is a promising and simple optimization technique. It has a single equation to be implemented, however, proper selection of the constants is very important.

10.6 GRAVITATIONAL SEARCH ALGORITHM

The Gravitational Search Algorithm (GSA) introduced in [129] follows the Newtonian gravity and laws of motion. Gravitation is the attraction of space bodies towards each other. Gravity is one of the fundamental force and the weakest of them all but is everywhere.

Some of the basic concepts of Newton's laws related to GSA are: (i) the acceleration of a particle is dependent on the force applied and its mass, (ii) effect of gravity will be more if the object applying gravity is bigger and closer to the body under observation and (iii) the increase in distance causes gravity to reduce.

There are different types of masses like active and passive gravitational mass, in the space, but in GSA, they are considered to be similar. The movement of the masses in the search space is dependent on the force experienced by the mass.

Eq 10.14 gives the expression for force experienced by mass 'i' due to mass 'j'.

$$F_{ij}^{d}(t) = G(t)\frac{M_{pi}(t) \times M_{aj}(t)}{R_{ij}(t) + \varepsilon}\left(x_j^d(t) - x_i^d(t)\right) \tag{10.14}$$

where 'M_{aj}' represents gravitational mass 'j' which is active, while 'M_{pi}' represents gravitational mass 'i' which is passive. 'ε' is considered a constant. 'x_j^d' represents the input variable value for j^{th} mass and d^{th} dimension while 'x_i^d' represents the same for i^{th} mass.

G(t) represents the gravitational constant at time 't' and is given in Eq 10.15.

$$G(t) = G(G_0, t) \tag{10.15}$$

The gravitational constant decreases with age of the universe and is represented through Eq 10.16, where 't_0' is the initial time.

$$G(t) = G(t_0) \times \left(\frac{t_0}{t}\right)^{\beta}, \quad \beta < 1 \tag{10.16}$$

In Eq 10.14, the distance between masses 'i' and 'j' is given by 'R_{ij}', which is the Euclidean distance and is represented by Eq 10.17.

$$R_{ij}(t) = \left\|X_i(t), X_j(t)\right\|_2 \tag{10.17}$$

To obtain the total force on a mass 'i' in d^{th} dimension due to combination of other masses, Eq 10.18 is used. However, this is not a normal summation of all forces from all masses. GSA involves, 'Kbest' agents for controlling the exploration and exploitation phenomena. The 'Kbest' value should decrease with time. Initial value of 'Kbest' is given by 'K_0', i.e., all masses shall apply force in the initial part of the search process, whereas it reduces linearly with increasing iterations, with one mass applying force on others at the end.

$$F_i^d(t) = \sum_{j \in Kbest, j \neq i} \text{rand}_j F_{ij}^d(t) \tag{10.18}$$

where '$rand_j$', is a random number between 0 and 1. Eq 10.18, represents total force exerted on mass 'i' in d^{th} dimension, as randomly weighted sum of dth components of 'Kbest' forces by 'Kbest' respective masses.

Thus, now by the law of motion, the acceleration of mass (also termed agent) at time 't' in d^{th} dimension is given by Eq 10.19.

$$a_i^d(t) = \frac{F_i^d(t)}{M_{ii}(t)} \tag{10.19}$$

where 'F_i' is the force experienced by mass 'i', given by Eq 10.14 and 'M_{ii}' is the internal mass of ith mass.

The velocity of the mass 'i' in d^{th} dimension is calculated as in Eq 10.20.

$$v_i^d(t+1) = rand_i \times v_i^d(t) + a_i^d(t) \tag{10.20}$$

In Eq 10.20, '$rand_i$' is a uniform random variable in the range of 0 and 1.

The new position of the mass is evaluated in Eq 10.21, which is similar to the method used in PSO.

$$x_i^d(t+1) = x_i^d(t) + v_i^d(t+1) \tag{10.21}$$

It is observed through above equations, that heavier masses move slowly and are more efficient.

The internal mass in Eq 10.19 can be evaluated through Eq 10.22.

$$M_i(t) = \frac{m_i(t)}{\sum_{j=1}^{N} m_j(t)} \tag{10.22}$$

where m_i is given through Eq 10.23

$$m_i(t) = \frac{fit_i(t) - worst(t)}{best(t) - worst(t)} \tag{10.23}$$

In Eq 10.23, fit_i gives the objective function value of mass 'i', whereas the expression for worst(t) and best(t) is given in Eq 10.24. Eq 10.24 is applicable for minimization problem. In a maximization problem, the min and max are interchanged in Eq 10.24.

$$\begin{aligned} \text{best}(t) &= \min_{j \in \{1,\ldots,N\}} fit_j(t) \\ \text{worst}(t) &= \max_{j \in \{1,\ldots,N\}} fit_j(t) \end{aligned} \tag{10.24}$$

GSA is a widely implemented optimization technique. However, as can be observed, a pitfall to this method is that there are too many equations to be applied. This can be assumed to contribute to the increase in time required for the execution process. On the other hand, the equations are simple which makes it easy to realise the optimization method.

10.7 REDUCING VARIABLE TREND SEARCH METHOD

Reducing Variable Trend Search (RVTS) was introduced in [11]. This optimization technique works on collective intelligence of the best solutions in the population. The optimal solution is reached when there is a consensus amongst the best solutions. In this method, based on the opinion of the selected best solutions, the search space is also reduced, which helps in speeding up the search process.

In RVTS, the candidate solutions are first randomly generated in the search space. The objective function of each solution is evaluated. The solutions are then sorted based on their objective function values. Next, a certain number of top solutions from the population are selected. These solutions are termed 'prime members'.

The word 'trend' is used in RVTS because the trend shown by the prime members is taken into consideration. To find the trend from the prime members, the values of each input variable of each prime member is varied from its minimum value to the maximum value. For each variation, the objective function value is monitored and the variable value for which minimum objective function value is obtained is stored. This process is presented in Fig 10.3.

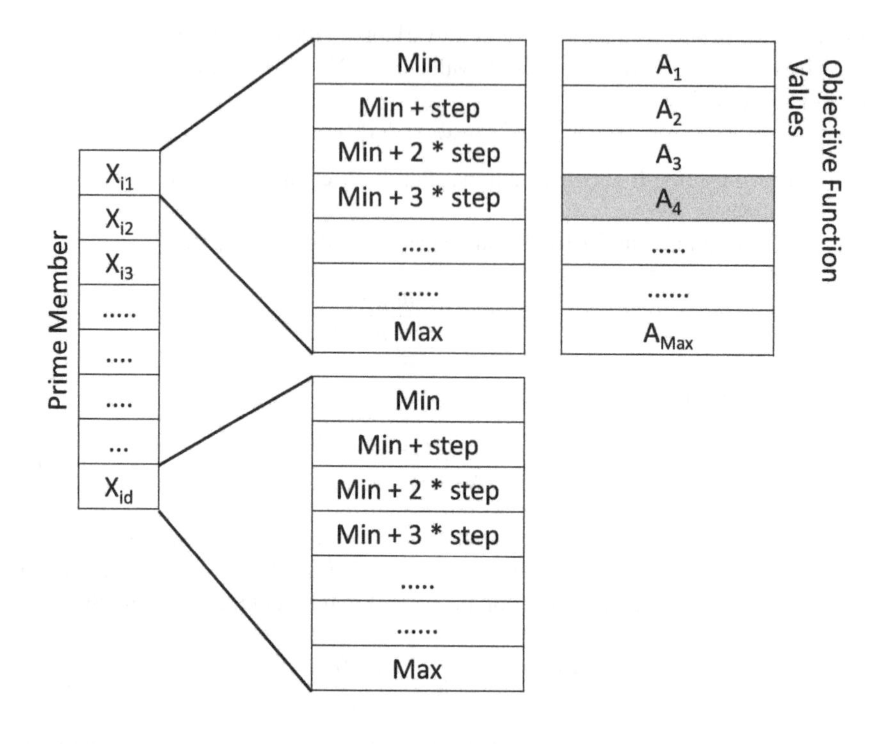

Figure 10.3: Trend Search

From Fig 10.3, it can be observed that for each prime member, the variable values are varied from minimum to maximum value and their objective function values are monitored. Thus for i^{th} prime member, shown in figure, the variable 1 is varied from Min to Max of its specified limits with an increment of a single step. This proces is repeated for each variable in each prime member. The minimum value of objective function obtained in this trend search is stored for each variable of each prime member in the MinSet matrix. In the figure, 'A_4' is shown as the minimum value for the trend search of first input variable of ith particle.

Thus, through this trend search, a certain trend can be obtained for the prime members. This trend is captured in a matrix termed MinSet matrix. The MinSet matrix is depicted in Fig 10.4. In the MinSet matrix, the values of input variables of prime members for which minimum objective function is obtained is stored.

In the figure, 'PM' represents Prime Member and 'IV' represents Input Variable. The values in the table are the input variable values corresponding to their respective trend search. Now for each row corresponding to the one input variable, the minimum and maximum values from the MinSet matrix are stored as new limits for the input variables. Thus for the next iteration, for input variable 1, the new limits would be 'IV_{1min}' to 'IV_{1max}'. The same is repeated for other input variables too. In the next generation, the solutions are again randomly generated, but within the new limits.

	PM_1	PM_2	PM_3	PM_4	.....	PM_N
IV_1	IV_{1max}	X	IV_{1min}	X		X
IV_2	IV_{2min}	X	X	X		IV_{2max}
IV_3	IV_{3min}	IV_{3max}	X	X		X
IV_4	X	X	IV_{4max}	IV_{4min}		X
....						
IV_d	IV_{dmax}	X	X	IV_{dmin}		X

Figure 10.4: MinSet Matrix

This causes the search space to reduce. There are some more conditions that are attached with the process involved.

The RVTS method brings with it some advantages as well as disadvantages. The solutions are generated again and again, thus avoiding the solutions being concentrated in a regional part of the search space. It is not dependent on a single global leader, but a number of solutions. The RVTS, however, cannot be applied easily to a continuous search space. Also the method is a little time consuming.

10.8 SUMMARY

This chapter presents a number of optimization techniques in brief. The optimization techniques are presented in the simplest way for the readers to understand and implement them. All the formulae related to the optimization techniques are also presented. The advantages and disadvantages of the methods are also expressed. The methods should be executed by the readers to get the feel of how the optimization techniques perform for different problems.

This chapter was added in the book to grab the opportunity to present multiple optimization techniques without going into detail but providing a brief insight about the workings of these techniques.

10.5 SUMMARY

Real-Time Application of PSO

In this book, a collection of Evolutionary Optimization Algorithms have been explained and demonstrated through a step-by-step solution. The real use of such algorithms can be understood through a practical application.

Thus, let us explore a simple application to view the real-time working of an optimization technique. The example contained in this section presents PSO in real action. Graphical user interface has been used for enhanced experience. The software used in this section is MATLAB® (2016).

Problem Statement: Reduce the losses occurring in the transmission system.

Objective: Reduce the flow of reactive power in the system. (Reactive Power is the power which does not perform any work. The presence of reactive power in the system will increase the losses occurring in the system. Thus to achieve the solution for minimization of losses, the reactive power within the transmission system should be reduced.)

Input: System Data, Input Variables and Constraints. (To find the losses occurring in the system, its details shall be required. The Newton Raphson or Gauss Seidel method can be employed to solve the load flow problem. Input variables shall be the generator voltages transformer taps and capacitor banks. The optimal combination of these variables shall reduce the reactive power in the system. There may be other input variables, too, but for the sake of simplicity we consider only these types of input variables. As far as constraints are considered, some constraints are inherent with the power flow, like, $P_g = P_d + P_l$, i.e., the total amount of generation should be equal to summation of total demand and total losses. Other constraints may include maximum and minimum allowable limits of bus voltages, etc.)

Test System: IEEE Standard 6 bus system

Figure 1 gives the IEEE 6 bus system:

The system details are given in Tables 1 and 2:

Table 1
Bus Data

1	**2**	**3**	**4**	**5**	**6**	**7**	**8**	9	10
1	1	1	0	0	0	0	0	0	0
2	2	1.1	0	0	0	50	0	−20	100
3	0	1	0	55	13	0	0	0	0
4	0	1	0	0	0	0	0	0	0
5	0	1	0	30	18	0	0	0	0
6	0	1	0	50	5	0	0	0	0

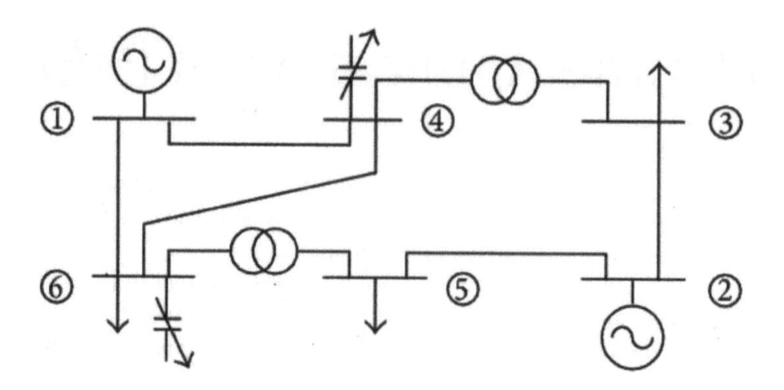

Figure 1: Standard IEEE 6 Bus System

Table 2
Line Data

From Bus	To Bus	Resistance	Reactance	Ratio
6	1	0.123	0.518	1
1	4	0.08	0.37	1
4	6	0.097	0.407	1
5	2	0.282	0.64	1
2	3	0.723	1.05	1
6	5	0	0.3	1.0725
4	3	0	0.133	1.06

The columns in Tables 1 and 2 are:
In Table 1:

- **1 - Bus No**: Gives the number of the bus in the system.
- **2 - Bus Type**: Specifies the bus type. Slack bus is given by 1, Generator/PV bus by 2 and Load/PQ bus by 0.
- **3 - Voltage Magnitude**: Gives the voltage magnitude to be assumed at a bus at the beginning of the load flow.
- **4 - Voltage Angle**: Gives the voltage angle to be assumed at a bus at the beginning of the load flow.
- **5 - Real Load**: Gives the real power load connected to a bus.
- **6 - Reactive Load**: Gives the reactive power load connected to a bus.
- **7 - Real Generation**: Gives the real power generation at a bus.
- **8 - Reactive Generation**: Gives the reactive power generation at a bus.
- **9 -** Q_{min}: Minimum allowable reactive power at a generation bus.
- **10 -** Q_{max}: Maximum allowable reactive power at a generation bus.

In Table 2:

From Bus: Bus No to which the element is connected to and considered as the start bus
To Bus: Bus No to which the element is connected to and considered as the end bus
Resistance: Resistance of the element
Reactance: Reactance of the element
Ratio: Transformer Tap Ratio in case the transformer is connected; otherwise it is taken as 1

Constraints: The constraints present in the system are 1) Power Balance constraint; as explained earlier, this constraint requires that the total generated power should be equal to sum of total load and total loss. 2) The permissible voltage range for any bus is from 0.9 p.u. to 1.1 p.u.

These two constraints will have to be satisfied during the evaluation of objective function value. The constraints which are to be satisfied through PSO are the limits of input variables. The input variables and their limits along with step sizes are given in Table 3.

Table 3
Input Variable Details

Input Variable	Min Limit	Max Limit	Step Size
V_1 (Voltage magnitude at Bus 1)	1	1.1	Continuous
V_2 (Voltage magnitude at Bus 2)	1.1	1.15	Continuous
Q_4 (Reactive Power injection at Bus 4)	0	5	0.5
Q_6 (Reactive Power injection at Bus 6)	0	5.5	0.5
T_{65} (Transformer Tap setting between Bus 6 and Bus 5)	0.91	1.11	0.0125
T_{43} (Transformer Tap setting between Bus 4 and Bus 3)	0.91	1.11	0.0125

Software: MATLAB 2016 above.

After gathering all the basic information related to the problem, let us get to designing of the application.

Application

In the application being designed, the PSO is supposed to find the optimal values of the input variables for different load conditions. The load is dynamic and can be changed during run time.

1. **Starting a new file**: Open MATLAB. In the Home tab, New → App → Guide, as shown in Fig 2. This will open the GUI interface model of MATLAB. A window pops up, which appears something like Fig 3.
2. **Setting up the Window**: The image of the system is placed as the background of the window. This is one of the ways to get the visual display of image on

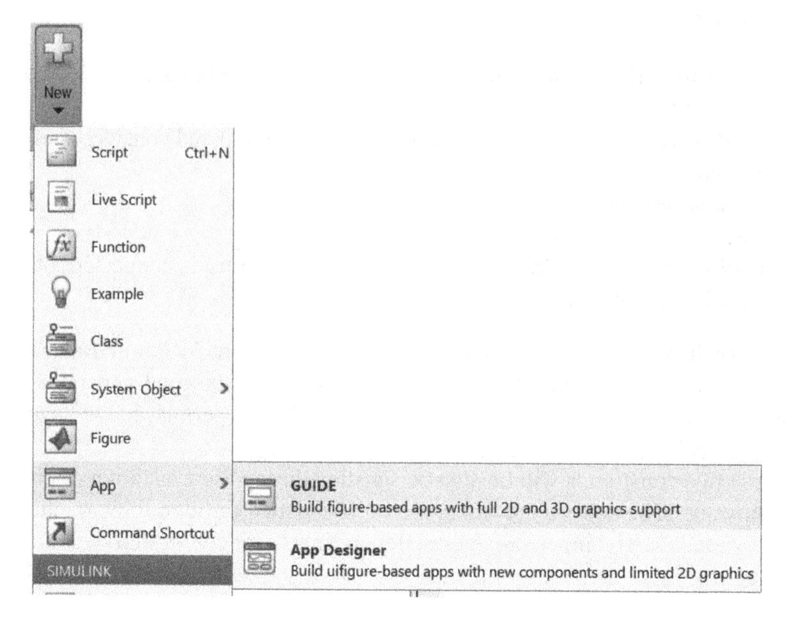

Figure 2: New GUI File

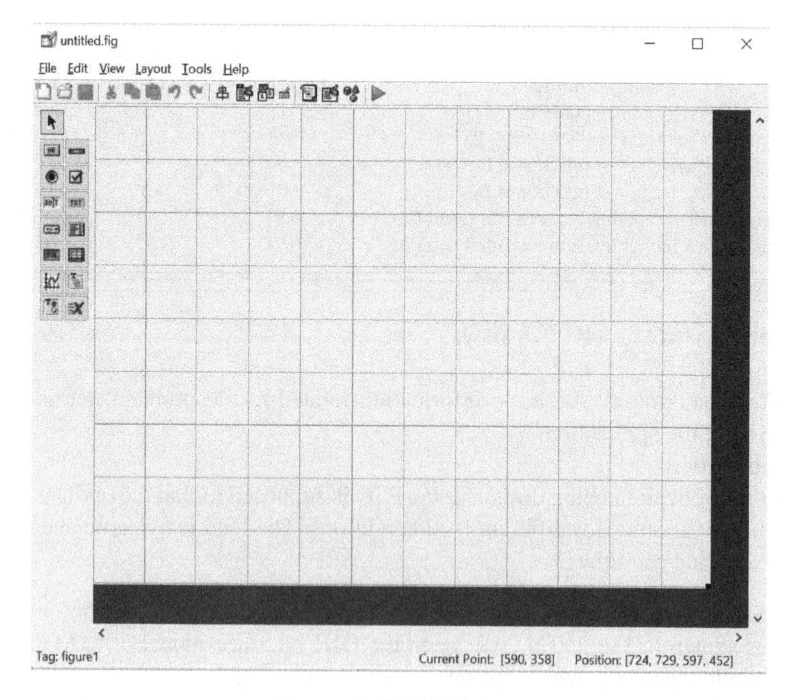

Figure 3: GUI Window

the window. The coding required to get the image in the background of the window is given in code below.

```
a=axes('unit','normalized','position',[0 0 1 1]);
b=imread('img.jpg'); imagesc(b);
set(a,'handlevisibility','off','visible','off')
```

where 1) 'axes' is used to create axes in the current figure, i.e., the app window. It will be used to position the image in the window. The first two arguments are used to size the axes according to the container, thus making the lower left corner as (0,0) and upper right corner as (1,1) of the figure. The last two arguments are used for placing the image. The first two arguments give the left and bottom values for positioning the image whereas the last two arguments give the width and height, respectively.
2) 'imread' function is used to read an image and store it in a variable. The image should be in the same folder where the GUI file is being stored, otherwise the whole path will have to be given.
3) 'imagesc' is used to display the image. In this case, the image had been coloured; it will display a colored image.
4) Set function is used to set some properties of axes. The properties being manipulated are handlevisibility and visible. Handlevisibility when set to 'off' makes the object handle invisible; i.e., in other words, no function shall be able to make changes in the axes. By setting visible to 'off', the object is hidden without getting deleted.
In all, this set of three commands is used to set the background for our application.
The output for the above code looks like Fig 4.

3. **Placing Controls on the Window**: Next the controls shall make entry onto the application window. From the Fig 4, it can be ascertained as to where the different buses, their loads and other elements are present in the window. Equipped with this the controls are drawn onto the window. The controls being drawn on the window are 'Text' and 'Slider'. The text controls will be required to display the optimal input variable values (6 nos), loss of the system (1 no) and the dynamic loads (3 nos). Slider controls shall be required (3 nos) to allow the user to change the load values. The input variables shall be optimized through PSO and displayed for each and every change in load values. The 'text' control is placed in the 3rd row, 2nd column and the 'slider' control is in 1st row, 2nd column of the panel in GUI window, refer Fig 3.
4. **Property Change of Controls**: The controls used in MATLAB GUI have properties which can be set through properties window. The properties window for each control can be accessed by double clicking the control. One can change a lot of properties, as can be viewed in Fig 5.
The first part of the image shows the properties of 'text' control and the second part show the properties for 'slider' control. The properties changed in 'text' control is FontSize to 12 and Tag and String are changed to 'T65'. These values shall be changed based on the input variable that it displays. For 'slider'

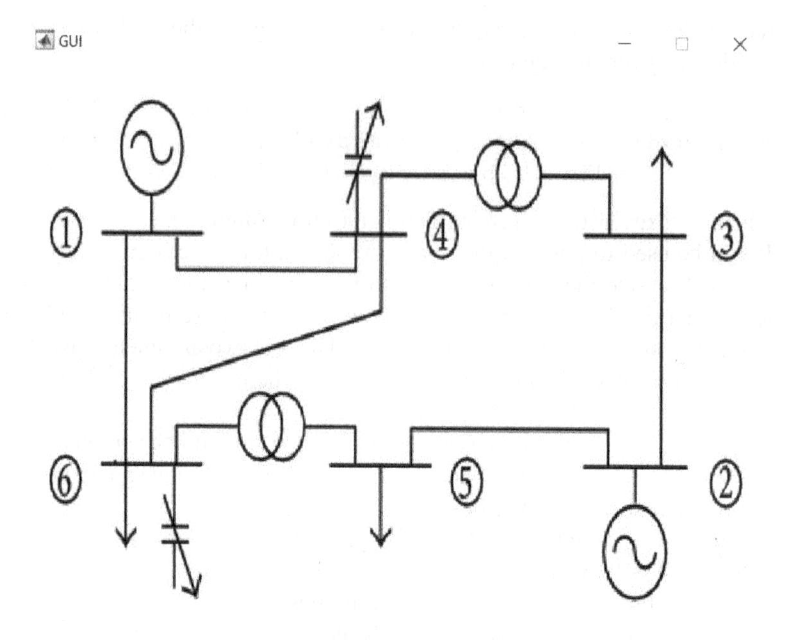

Figure 4: GUI with Background

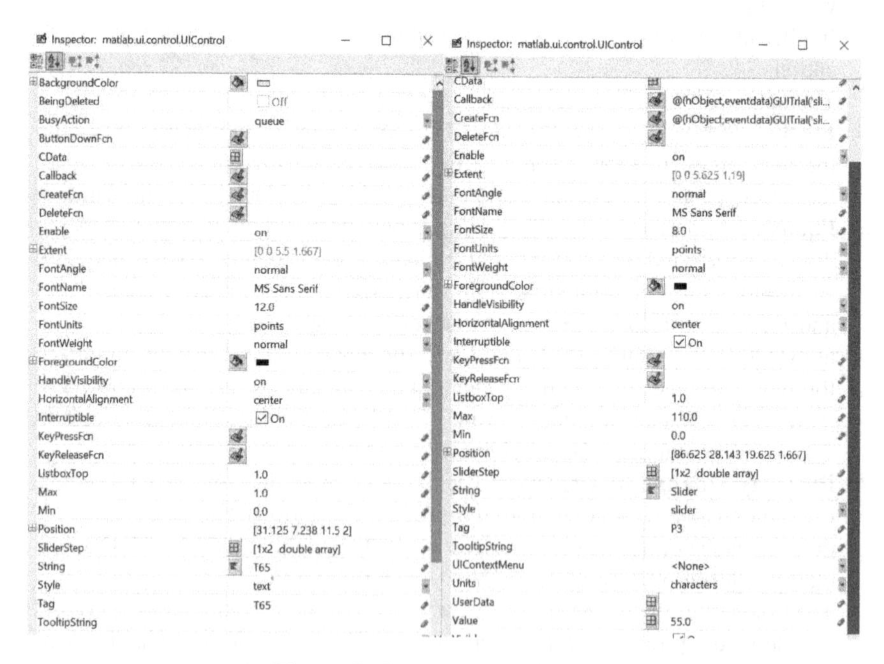

Figure 5: Control Properties in GUI

control the properties changed are: (i) Max – double of standard load, like if the standard load at the said bus was 55, then value of Max will be 110, (ii) SliderStep - changed to 0.1 and 0.3, these values represent the changes in the slider control during run time, in which 0.1 represents 10% whereas 0.3 represents 30% change in slider value when clicked on arrow and inside the control, respectively, (iii) Value - changed to standard mentioned value in the system, and (iv) Tag – changed to 'P3', to denote that the slider control is placed at bus 3. For other buses, other values should be used. None of the properties changed in the 'slider' control shall be visible at run time; however, its effects can be observed. All the 'text' and 'slider' controls are placed on the window and are shown in Fig 6. After running the GUI, it appears as in Fig 7.

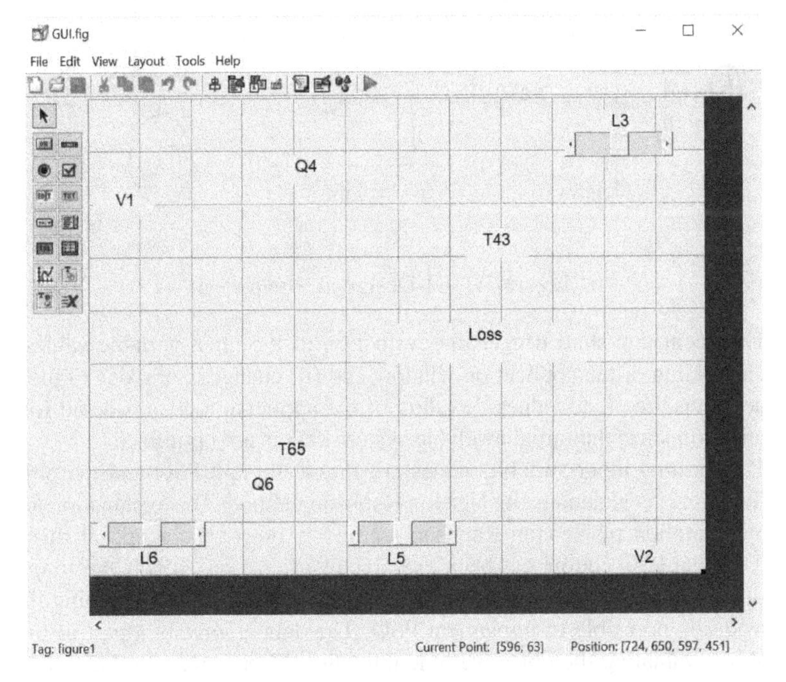

Figure 6: Control Placement on GUI

5. **Coding**: Now that the initial setting is done for the application, we need to insert the code at the right place. The MATLAB GUI coding is event based and any action taken by the user on the application window is considered an event. These events could be clicking of a button, change in slider position, loading of application, etc.
   The aim of the application is to obtain optimal combination of input variables to reduce active power losses in real time. Thus, the optimal combination of input variables should be available for every change in slider position as done by the user of the application.

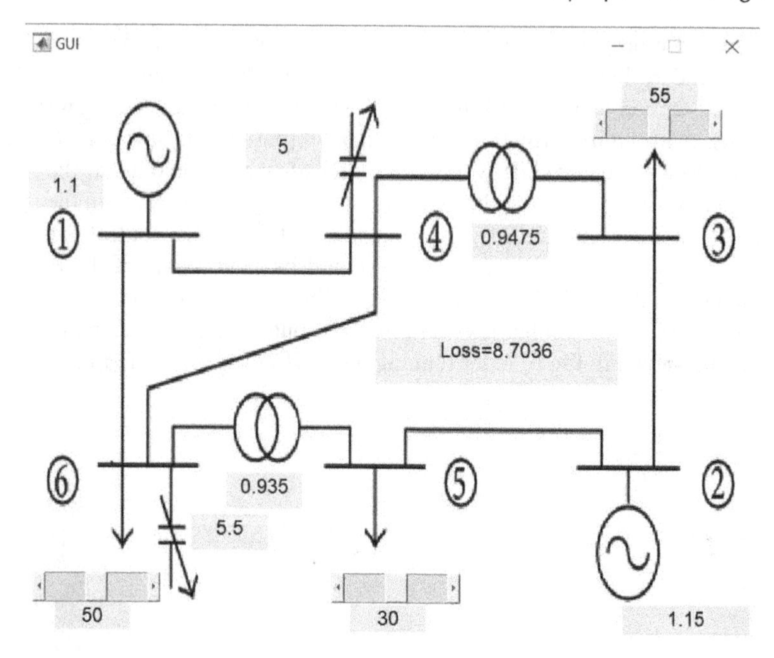

Figure 7: GUI-Designed Window

The event that shall trigger the evaluation of the input variable values is (i) the loading of the application window and (ii) change in any one of the slider positions, i.e., load values. Loading of the application is considered for optimization since the initial available system data is not optimized.
The optimal input variable values are found through PSO and the objective function is evaluated using Newton Raphson method. The explanation and implementation of Newton Raphson method is out of the scope of this book. Also, the PSO coding has been explained earlier. Thus, it is better to only visualize the code required to execute this application while assuming that the reader is now able to implement PSO. The data is already a part of the program evaluating the objective function value. Therefore, only the new values of slider need to be sent to represent the changed load values at different buses. The code below gives the required statements:

```
P6=handles.P6.Value;
P5=handles.P5.Value;
P3=handles.P3.Value;
res=PSO(P3,P5,P6);
handles.V1.String=res(1);
handles.V2.String=res(2);
handles.Q4.String=res(3);
handles.Q6.String=res(4);
handles.T65.String=res(5);
handles.T43.String=res(6);
```

```
handles.Loss.String=strcat('Loss=',
num2str(res(7)));
handles.P6t.String=handles.P6.Value;
handles.P5t.String=handles.P5.Value;
handles.P3t.String=handles.P3.Value;
```

GUI mechanism has a handles structure for storing and retrieving of shared data. In simple words, 'handles' is used to retrieve and change properties of GUI controls.

The code can be explained in three parts: first is to read the change in load values, second is to perform optimization of the system and lastly to display the values. In the first three lines, the load values are read from the slider controls. The slider controls represent the load value at different load buses. Value property is read from the control and stored in variables 'P6', 'P5' and 'P3'. The new load values are then sent to the PSO function to obtain the optimal combination of input variables. The PSO function uses PSO optimization technique and Newton Raphson method to output the required results which is stored in 'res' variable. The 'res' variable contains 6 optimized input variable values and the loss. These values are then transferred in different 'text' controls through their 'String' property. The last three lines of the code transfers the 'slider' control value onto the 'String' property of the 'text' controls.

6. **Output**: Let us observe different outputs obtained for changes performed on load values shown in Figs 8 – 13. The system operation for different load

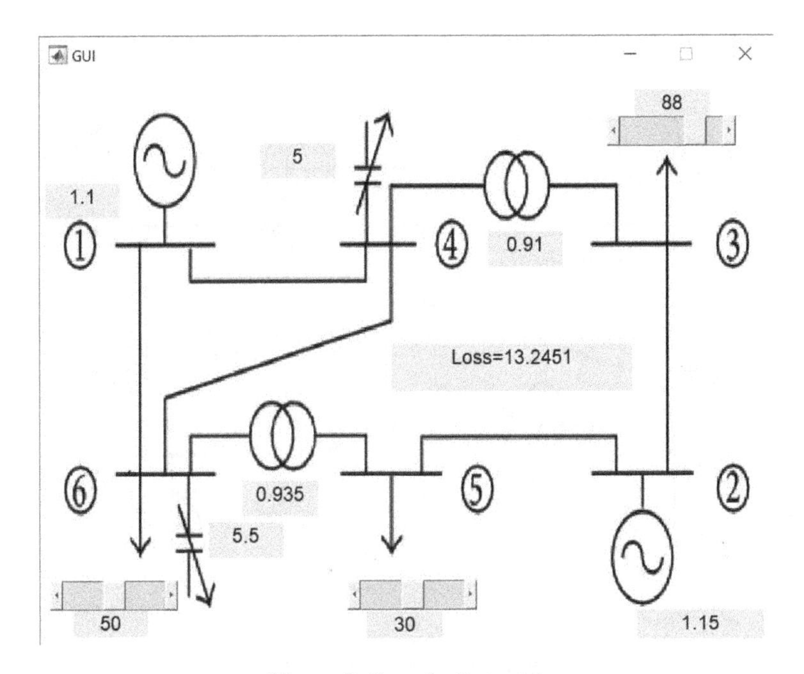

Figure 8: Sample Output 1

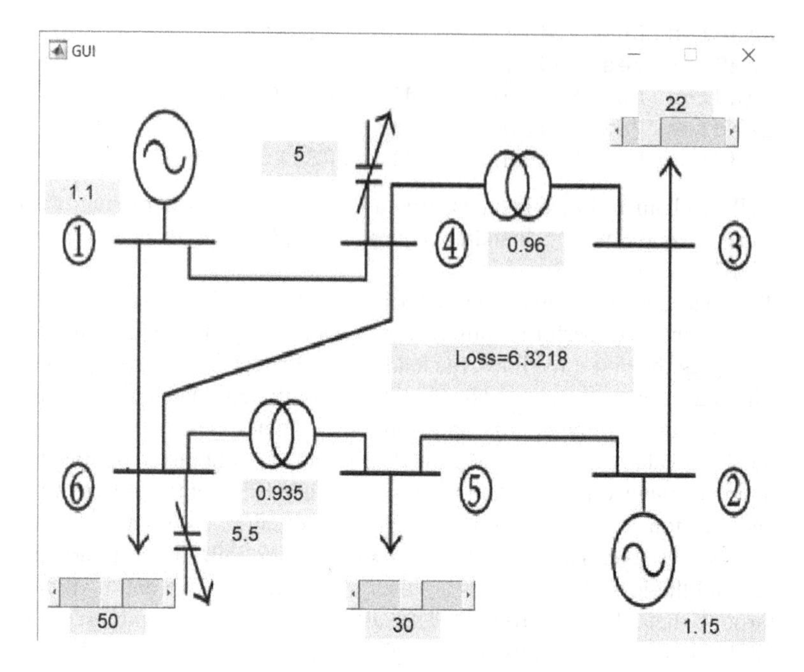

Figure 9: Sample Output 2

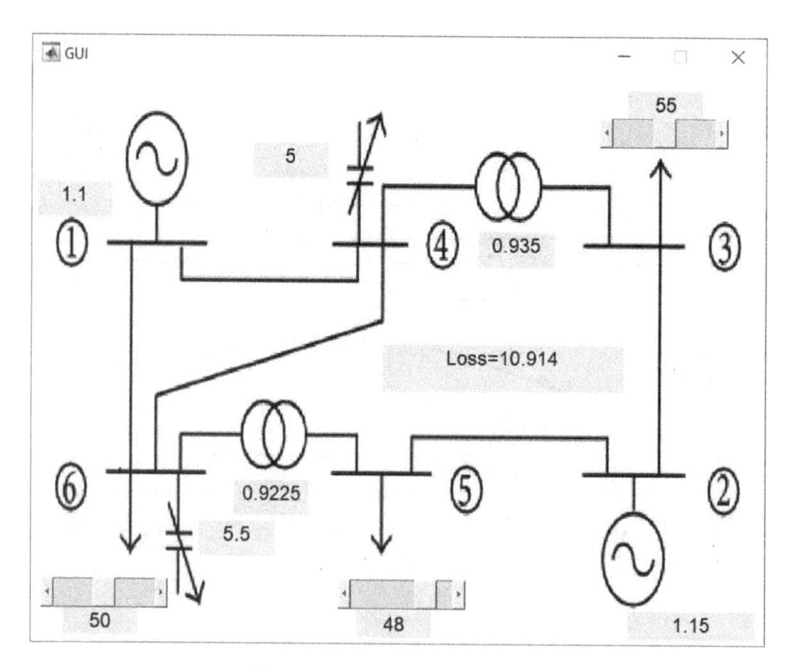

Figure 10: Sample Output 3

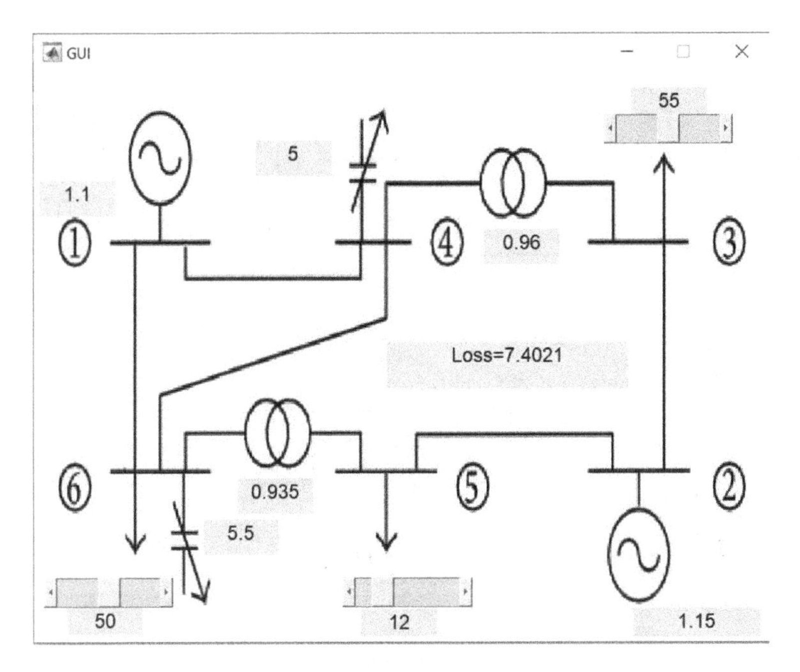

Figure 11: Sample Output 4

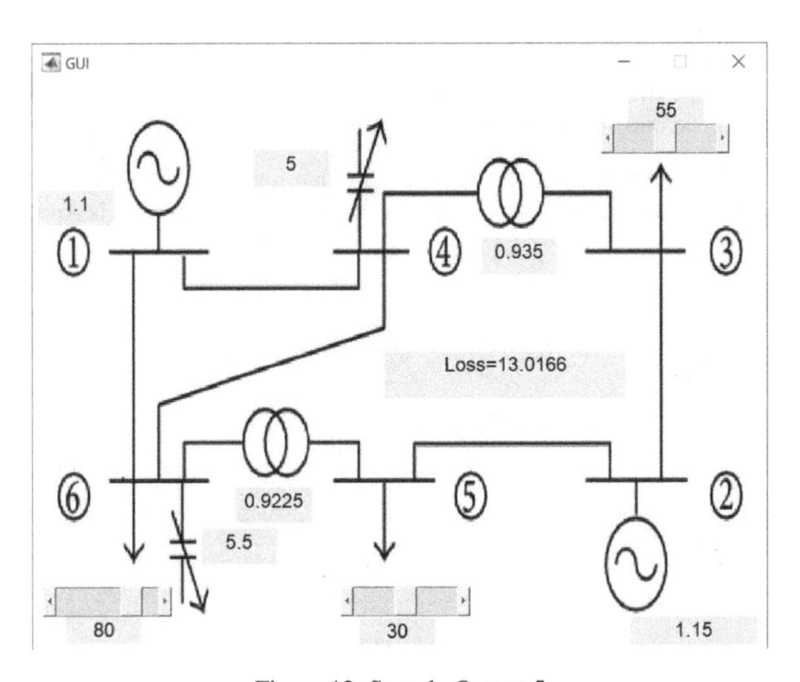

Figure 12: Sample Output 5

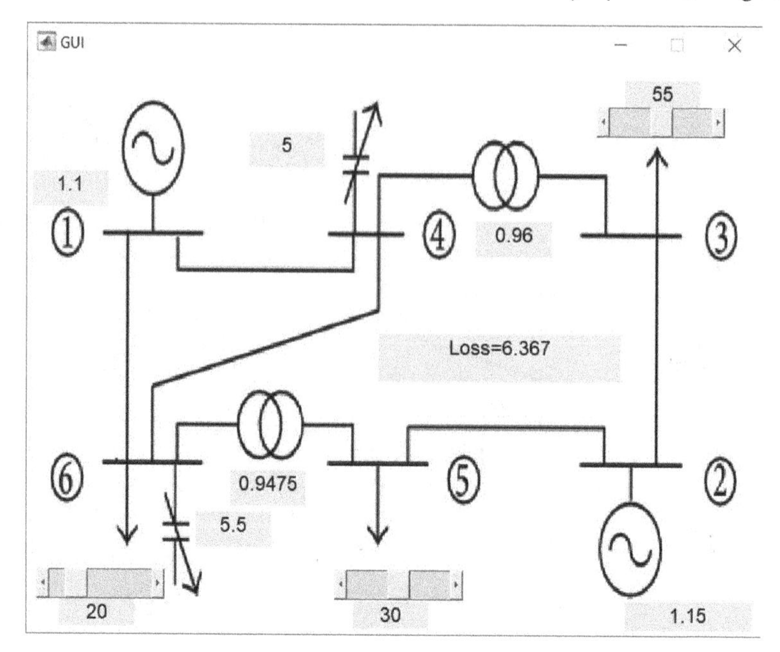

Figure 13: Sample Output 6

conditions can be observed. More complex situations and circuits can also be implemented to observe their effects on the input variable values. Also, instead of having slider controls, the load data can be read through live online systems or can be imported from a forecasting model.

The example should help the users to start their own GUI real-time application development and experience the different features available through this Appendix.

Optimization Techniques in Python

In this section of the book, the programming language of Python is introduced to solve optimization technique problems. The objective of this Appendix is introduction to different optimization-related packages in Python. One has to remember that: (i) for mastering the packages, practice is required and (ii) with time newer versions and newer packages shall keep on getting introduced.

Python is a very promising language. It has been growing at a very fast pace. It is free and has contribution from a wide variety of contributors. Python helps you integrate easily and effectively. Python has been widely implemented in different areas. To download Python visit: www.python.org. One can use Jupyter as its editor, which is available at: https://jupyter.org/, otherwise "Idle" is provided along with the Python installation package. Some of the scientific packages for Python currently include: SciPy, Pandas, IPython.

Python was first released in 1991 by Guido Van Rossum. Python supports structured programming, OOPS and functional programming. Knowledge of C and OOPS should be enough for someone to start with Python programming. Python has grown a lot and thus the scope of this section is limited to "application of Python to optimization problems". We will view the usage of different packages/modules available in Python for implementing optimization techniques. To experience these packages, one has to set the path for the scripts folder in Python39, or go to the scripts folder in command prompt and then install the package. We will try to explore some packages henceforth.

1) **Genetic Algorithm**

As mentioned before, go to the scripts folder or set path for the folder. Open the command prompt and type "pip install geneticalgorithm". The package belongs to PyPi repository.

One can see the package getting installed in Python. Some other packages required for successful running of 'geneticalgorithm' package are numpy and matplotlib. So, these packages are also required to be installed.

Once the package has been installed, let us open a new file. Inside the file, our objective is to solve the sphere function through genetic algorithm package for 5 input variables.

The approximate code is already present on the website:

https://pypi.org/project/geneticalgorithm/

This code can be classified into three parts: declaring of variables, defining of objective function and utilizing the genetic algorithm package.

The code for declaring of variables is:

```
import numpy as n
from geneticalgorithm import geneticalgorithm as g
```

A variable ‘n’ is created, with a datatype of numpy. ‘numpy’ package is of great use for array implementation and in the field of linear algebra, matrices, FFT, etc. The next variable created is ‘g’ for implementing the geneticalgorithm package. One should understand the different methods of declaring variables in Python before starting to write programs in it.

In the next part of the code, we need to define the objective function. Since we want to implement the sphere function, the code would be:

```
def sph(X):
return n.sum(X**2)
```

The name of the function defined is ‘sph’, which receives ‘X’ as the argument and returns the sum of squares of elements in ‘X’. To find the square the symbol used is ‘**’. ‘sum’ is a method under ‘n’, i.e., numpy.

In the last part of the program, the code used should be

```
limits=n.array([[0,5.12]]*5)
m=g(function=sph,dimension=5,variable_type='real',...
variable_boundaries=limits)
m.run()
```

In this part, an array is first generated through the numpy variable ‘n’. ‘array’ function is used for the same. The command shall generate a matrix of size 5 × 2 with each value in the first column being 0 and that in the second column being 5.12, as shown below:

0	5.12
0	5.12
0	5.12
0	5.12
0	5.12

This will ensure that the input variables shall be generated within the given range/limits. In case, the limits vary for input variables, then the limits specified in each row can be arranged accordingly. The next command is used to set up a model for genetic algorithm. The arguments required for setting up this model are

function: gives the name of the function that finds the objective function. In our case, it is “sph”

dimension: number of input variables. In our case, it is 5

variable_type: data type of the variable. In our case, “real” is chosen, which means that decimal numbers are selected. Other datatypes which can be used are “int” or “bool”. In case, the input variables have different datatypes, then a matrix is required to be generated specifying their datatypes, respectively.

variable_boundaries: this argument gives the limits of the input variables. The limit matrix is already created in the earlier command for this purpose.

In the last command, the model created is executed through the "run" command.

Some of the other parameters, that can be set for the genetic algorithm are: max_num_iteration, population_size, mutation_probability, crossover_probabilty, crossover_type, elit_ratio, parents_portion and max_iteration_without_improv. To set these values, the code used would be:

```
alg_param = {'max_num_iteration':␣3000,...
'population_size':100,'mutation_probability':0.1,...
'elit_ratio': 0.01,'crossover_probability':␣0.5,...
'parents_portion': 0.3,'crossover_type':'uniform',...
'max_iteration_without_improv':None}
```

Many of these parameters can be understood if one goes through the genetic algorithm chapter of this book. Still, for the benefit of the readers, they are explained in brief again.

1. **max_num_iteration**: It specifies the maximum number of iterations. If not specified then it will be set based on dimension, boundaries and population.
2. **population_size**: It gives the number of chromosomes. Default value is 100.
3. **mutation_probability**: Gives the probability of a gene in a chromosome getting replaced by a random value. Default value for this argument is 0.1 (i.e., 10 percent).
4. **crossover_probability**: Gives the probability of an existing parent to transfer its characteristics to a child chromosome. Default argument value is taken as 0.5 (i.e., 50 percent).
5. **crossover_type**: Specifies the methodology used in crossover process. Available options are one_point, two_point, and uniform crossover. Default value of crossover_type is uniform.
6. **elit_ratio**: Specifies the number of parents to be passed over to the next generation as elite. Default value is considered 0.01 (i.e., 1 percent).
7. **parents_portion**: Number of parents from previous generation saved to next generation including elit_ratio. Default is considered 0.3 (i.e., 30 percent of population).
8. **max_iteration_without_improv**: This parameter is used to stop the optimization process if the objective function value does not improve for the specified number of successive iterations. Default value is taken as None.

After setting of the algorithm parameter, the model can be executed as:

```
m=g(function=sph,dimension=5,variable_type='real',...
variable_boundaries=limits,
algorithm_parameters=alg_param)
```

The variations that are introduced should be carefully set for faster and efficient execution of the optimization technique.

With the default setting of the parameters the genetic algorithm shall take a long time for execution. The best solution found after execution of the code was:

[0.00020718 0.00522644 0.00780962 0.01123254 0.01213513]

And the objective function value is: 0.0003617801995816604

A geneticalgorithm2 package is also available at https://pypi.org/project/ geneticalgorithm2/. This package is similar to the one explained and can be explored further.

2) **pyswarm**: This package currently has a single function of "pso". It can be installed through "pip" command and is available at https://pythonhosted.org/pyswarm/. The prerequisite for this package is "numpy".

For this package as well, optimization of sphere function with 5 input variables will be performed. The code shall be divided into three parts again: declaration, defining objective function, execution of PSO.

In the initial part, the declaration for the required packages is given as:

```
import numpy as n
from pyswarm import pso
```

The numpy variable is not a necessity if one is not using the functions associated with it.

In the next part, the definition of objective function and limits is performed as:

```
def sph(x):
    return n.sum(x**2)

ll = [0, 0, 0, 0, 0]
ul = [5.12, 5.12, 5.12, 5.12, 5.12]
```

The "sph" function defines the evaluation of objective function value. The objective function should be for minimization. 'll' is an array of the lower limits for the input variables whereas 'ul' provides the upper limits for these input variables, respectively.

In the last part of the code, the "pso" function is executed. The arguments required to be passed are: objective function name, lower limit array and upper limit array.

```
xopt , fopt = pso(sph , ll, ul)
```

'xopt' will contain the optimal input variable values and 'fopt' shall have the optimal objective function value.

The execution of the function is very fast and it is quite easy to encode. This is the minimum version of "pso" function. Next, let us explore some more extensions associated with it. These extensions can be added in the call for "pso" function.

1. **f_ieqcons**: It is possible to define a function for generating constraints. The function is defined in a similar way as the objective function, but is expected to return an array of values. In the "pso" function call, the parameter used for addressing the constraint is 'f_ieqcons'. Default is taken as None

2. **ieqcons**: It provides for a list of functions. Default is an empty list.
3. **swarmsize**: Specifies the population size.
4. **omega**: Inertia weight, default is taken as 0.5. Suggested to be specified between 0 and 1.
5. **phip**: Weighing factor for individual best, default is 0.5. Suggested to be specified between 0 and 1.
6. **phig**: Weighing factor for global best, default is 0.5. Suggested to be specified between 0 and 1.
7. **maxiter**: Maximum number of iterations, default is 100.
8. **minstep**: This is a stopping criterion to check the minimum improvement in global best position, default is taken as 1e-8.
9. **minfunc**: This is a stopping criterion to check the minimum improvement in global best objective function, default is taken as 1e-8.
10. **debug**: It is a Boolean argument. If specified as True, then progress statement for each iteration shall be displayed. Default value is False.
11. **args**: 'args' used to supply some extra settings or information to the objective or constraint functions. Default is empty.
12. **kwargs**: Additional keywords arguments for objective or constraint functions. Default is empty.

Some of the other packages that can be explored are:

1. **Evolutionary Optimization**: It deals with a number of optimization techniques. Available at: https://pypi.org/project/EvoOpt/.
2. **sklearn-genetic**: This package contains a lot of parameters and settings for genetic algorithm. Available at: https://pypi.org/project/sklearn-genetic/.
3. **inspyred**: There are a lot of options possible in this package. It presents its own set of functions for specific tasks. Available at: https://pythonhosted.org/inspyred/.
4. **pyeasyga**: This is another package for implementing GA. Available at: https://pypi.org/project/pyeasyga/.
5. **geneal**: This is another version for implementing genetic algorithm. Available at: https://towardsdatascience.com/introducing-geneal-a-genetic-algorithm-python-library-db69abfc212c.
6. **scipy.optimize.differential_evolution**: This package offers to implement Differential Evolution algorithm. It offers plenty of parameters that can be used to fine tune the operation of the algorithm. Available at: https://docs.scipy.org/doc/scipy /reference/generated/scipy.optimize.differential_evolution.html.
7. **Distributed Evolutionary Algorithms in Python (DEAP)**: DEAP is a very advanced package consisting of parallel mechanism, rapid prototyping, etc. Even though it is more oriented towards genetic algorithm, the documentation mentions the presence of other optimization techniques also. Multiobjective optimization can also be implemented. Available at: https://pypi.org/project/deap/.
8. **Pyvolution**: This is an Evolutionary algorithms framework. Available at: https://pypi.org/project/Pyvolution/.

9. **geneticalgs**: Another package dealing with genetic algorithms. Available at: https://pypi.org/project/geneticalgs/.
10. **ga2**: One more genetic algorithm inspired package. Available at: https://pypi.org/project/ga2/.
11. **Discrete Genetic Algorithm (dga)**: This package implements discrete genetic algorithm. Available at: https://pypi.org/project/dga/.
12. **pygalib**: Another genetic algorithm resource. Available at: https://pypi.org/project /pygalib/.
13. **pymoo**: A comprehensively developed package for multiobjective optimization along with examples. Available at: https://pymoo.org/.
14. **Platypus**: Another package for multiobjective optimization. Available at: https://platypus.readthedocs.io/en/latest/index.html#.

The above list is a collection of different available packages in Python for implementing evolutionary optimization techniques. With time, may be their websites might change, their versions may change or may be the packages may no longer be present at the given websites. Also, there may be many more packages available online, and the reader may search the Internet space. But, it is suggested that in case one is developing his/her own project, then it is better to develop your own code rather than depending on some code which may not be according to the project's requirement.

Standard Optimization Problems

A collection of standard optimization problems used in different literature is presented in this section of the Appendix. The optimization problems dealing with a single input variable are not listed in this table.

Table 4
Standard Optimization Problems

Sr. No.	Function Name	Dimensions	Range	Equation
1	Bukin N. 6	2	$x_1 \in [-15, -5]$, $x_2 \in [-3, 3]$	$100\sqrt{\lvert x_2 - 0.01x_1^2\rvert} + 0.01\lvert x_1 + 10\rvert$
2	Cross-In-Tray	2	[−10, 10]	$-0.0001\left(\left\lvert \sin(x_1)\sin(x_2)\exp\left(\left\lvert 100 - \frac{\sqrt{x_1^2+x_2^2}}{\pi}\right\rvert\right)\right\rvert + 1\right)^{0.1}$
3	Stepint	5	[−5.12, 5.12]	$25 + \sum_{i=1}^{5} \lfloor x_i \rfloor$
4	Step	30	[−100, 100]	$\sum_{i=1}^{n} (\lfloor x_i + 0.5 \rfloor)^2$
5	SumSquares	30	[−10, 10]	$\sum_{i=1}^{n} i x_i^2$
6	Quartic	30	[−1.28, 1.28]	$\sum_{i=1}^{n} i x_i^4 + random(0,1)$
7	Beale	5	[−4.5, 4.5]	$(1.5 - x_1 + x_1x_2)^2 + (2.25 - x_1 + x_1x_2^2)^2 + (2.625 - x_1 + x_1x_2^3)^2$
8	Easom	2	[−100, 100]	$-cos(x_1)cos(x_2)exp(-(x_1-\pi)^2 - (x_2-\pi)^2)$
9	Langerman2	2	[0 10]	$-\sum_{i=1}^{m} c_i(exp(-\frac{1}{\pi}\sum_{j=1}^{n}(x_j - a_{ij})^2 cos(\pi\sum_{j=1}^{n}(x_j - a_{ij})^2))$
10	Matyas	2	[−10 10]	$0.26\,(x_1^2 + x_2^2) - 0.48x_1x_2$
11	Colville	4	[−10 10]	$100\,(x_1^2 - x_2)^2 + (x_1-1)^2 + (x_3-1)^2 + 90\,(x_3^2 - x_4)^2 + 10.1\left((x_2-1)^2 + (x_4-1)^2\right) + 19.8\,(x_2-1)\,(x_4-1)$
12	Trid6	6	$[-D^2\ D^2]$	$\sum_{i=1}^{n}(x_i-1)^2 - \sum_{i=2}^{n} x_i x_{i-1}$
13	Powell	24	[−4 5]	$\sum_{i=1}^{n/k}(x_{4i-3} + 10x_{4i-2})^2 + 5\,(x_{4i-1} - x_{4i})^2 + (x_{4i-2} - x_{4i-1})^4 + 10\,(x_{4i-3} - x_{4i})^4$
14	Schwefel 2.22	30	[−10 10]	$\sum_{i=1}^{n}\lvert x_i\rvert + \prod_{i=1}^{n}\lvert x_i\rvert$
15	Schwefel 1.2	30	[−100 100]	$\sum_{i=1}^{n}\left(\sum_{j=1}^{i} x_j\right)^2$
16	Dixon-Price	30	[−10 10]	$(x_1-1)^2 + \sum_{i=2}^{n} i\,(2x_i^2 - x_{i-1})^2$
17	Foxholes	2	[−65.536 65.536]	$\left[\frac{1}{500} + \sum_{j=1}^{25}\frac{1}{j+\sum_{i=1}^{2}(x_i - a_{ij})6}\right]^{-1}$
18	Branin	2	[5, 10]x[0,15]	$\left(x_2 - \frac{5.1}{4\pi^2}x_1^2 + \frac{5}{\pi}x_1 - 6\right)^2 + 10\left(1 - \frac{1}{8\pi}\right)\cos x_1 + 10$
19	Bohachevsky1	2	[−100 100]	$x_1^2 + 2x_2^2 - 0.3\cos(3\pi x_1) - 0.4\cos(4\pi x_2) + 0.7$
20	Booth	2	[−10 10]	$(x_1 + 2x_2 - 7)^2 + (2x_1 + x_2 - 5)^2$
21	Schwefel	30	[−500 500]	$\sum_{i=1}^{n} -x_i \sin\left(\sqrt{\lvert x_i\rvert}\right)$
22	Michalewicz2	2	$[0\ \pi]$	$-\sum_{i=1}^{n}\sin(x_i)\left(\sin(ix_i^2/\pi)\right)^{2m}$ where m=10

Sr. No.	Function Name	Dimensions	Range	Equation
23	Schaffer	2	[−100 100]	$0.5+\frac{\sin^2\left(\sqrt{x_1^2+x_2^2}\right)-0.5}{\left(1+0.001\left(x_1^2+x_2^2\right)\right)^2}$
24	Six Hump Camel Back	2	[−5 5]	$4x_1^2-2.1x_1^4+\frac{1}{3}x_1^6+x1x2-4x_2^2+4x_2^4$
25	Bohachevsky2	2	[−100 100]	$x_1^2+2x_2^2-0.3\cos(3\pi x_1)(4\pi x_2)+0.3$
26	Bohachevsky3	2	[−100 100]	$x_1^2+2x_2^2-0.3\cos(3\pi x_1+4\pi x_2)+0.3$
27	Shubert	2	[−10 10]	$\left(\sum_{i=1}^{5} i\cos((i+1)x_1+i)\right)\left(\sum_{i=1}^{5} i\cos((i+1)x_2+i)\right)$
28	Kowalik	4	[−5 5]	$\sum_{i=1}^{11}\left(a_i-\frac{x_1\left(b_i^2+b_ix_2\right)}{b_1^2+b_ix_3+x_4}\right)^2$
29	ShekelN	4	[0 10]	$-\sum_{i=1}^{N}\left[(x-a_i)(x-a_i)^T+c_i\right]^{-1}$
30	Perm	4	[−D D]	$\sum_{k=1}^{n}\left[\sum_{i=1}^{n}\left(i^k+\beta\right)\left((x_i/i)^k-1\right)\right]^2$
31	PowerSum	4	[0 D]	$\sum_{k=1}^{n}\left[\left(\sum_{i=1}^{n}x_i^k\right)-b_k\right]^2$
32	HartmanN	N	[0 1]	$-\sum_{i=1}^{4}c_i\exp\left[-\sum_{j=1}^{N}a_{ij}\left(x_j-p_{ij}\right)^2\right]$
33	Drop-Wave	2	[−5.12, 5.12]	$-\frac{1+\cos\left(12\sqrt{x_1^2+x_2^2}\right)}{0.5\left(x_1^2+x_2^2\right)+2}$
34	Eggholder	2	[−512, 512]	$-(x_2+47)\sin\left(\sqrt{\lvert x_2+\frac{x_1}{2}+47\rvert}\right)-x_1\sin\left(\sqrt{\lvert x_1-(x_2+47)\rvert}\right)$
35	Holder Table	2	[−10, 10]	$-\left\lvert\sin(x_1)\cos(x_2)\exp\left(\left\lvert 1-\frac{\sqrt{x_1^2+x_2^2}}{\pi}\right\rvert\right)\right\rvert$
36	Levy N. 13	2	[−10, 10]	$\sin^2(3\pi x_1)+(x_1-1)^2\left[1+\sin^2(3\pi x_2)\right]+(x_2-1)^2\left[1+\sin^2(2\pi x_2)\right]$
37	Styblinski-Tang	d	[−5, 5]	$\frac{1}{2}\sum_{i=1}^{d}\left(x_i^4-16x_i^2+5x_i\right)$
38	Rotated Hyper-Ellipsoid	d	[−65.536, 65.536]	$\sum_{i=1}^{d}\sum_{j=1}^{i}x_j^2$
39	Sum Of Different Powers	d	[−1, 1]	$\sum_{i=1}^{d}\lvert x_i\rvert^{i+1}$
40	Booth	2	[−10, 10]	$(x_1+2x_2-7)^2+(2x_1+x_2-5)^2$
41	McCormick	2	$x_1\in[-1.5,4], x2\in[-3,4]$	$\sin(x_1+x_2)+(x_1-x_2)^2-1.5x_1+2.5x_2+1$
42	Three-Hump Camel	2	[−5, 5]	$2x_1^2-1.05x_1^4+\frac{x_1^6}{6}+x_1x_2+x_2^2$

Sr. No.	Function Name	Dimensions	Range	Equation
43	Levy	d	[−10, 10]	$\sin^2(\pi w_1)+\sum_{i=1}^{d-1}(w_i-1)^2\left[1+10\sin^2(\pi w_i+1)\right]+(w_d-1)^2\left[1+\sin^2(2\pi w_d)\right]$, where $w_i=1+\frac{x_i-1}{4}$, for all $i=1,\ldots,d$
44	De Jong N. 5	2	[−65.536, 65.536]	$\left(0.002+\sum_{i=1}^{25}\frac{1}{i+(x_1-a_{1i})^6+(x_2-a_{2i})^6}\right)^{-1}$, where $\mathbf{a}=\begin{pmatrix} -32 & -16 & 0 & 16 & 32 & -32 & \ldots & 0 & 16 & 32 \\ -32 & -32 & -32 & -32 & -32 & -16 & \ldots & 32 & 32 & 32 \end{pmatrix}$
45	Goldstein-Price	2	[−2, 2]	$\left[1+(x_1+x_2+1)^2(19-14x_1+3x_1^2-14x_2+6x_1x_2+3x_2^2)\right]\times\left[30+(2x_1-3x_2)^2(18-32x_1+12x_1^2+48x_2-36x_1x_2+27x_2^2)\right]$

Bibliography

1. Traveling salesman problems with constraints: the TSP with time windows. https://acrogenesis.com/or-tools/documentation/user_manual/manual/TSP.html. Accessed: 2020-01-06.

2. Hussein A Abbass. The self-adaptive pareto differential evolution algorithm. In *Proceedings of the 2002 Congress on Evolutionary Computation. CEC'02 (Cat. No. 02TH8600)*, volume 1, pages 831–836. IEEE, 2002.

3. Malik Muneeb Abid and Iqbal Muhammad. Heuristic approaches to solve traveling salesman problem. *Indonesian Journal of Electrical Engineering and Computer Science*, 15(2):390–396, 2015.

4. MA Abido. Optimal Power Flow using Particle Swarm Optimization. *International Journal of Electrical Power & Energy Systems*, 24(7):563–571, 2002.

5. MA Abido. Multiobjective particle swarm optimization for environmental/economic dispatch problem. *Electric Power Systems Research*, 79(7):1105–1113, 2009.

6. Ernesto P Adorio and U Diliman. MVF-multivariate test functions library in C for unconstrained global optimization. *Quezon City, Metro Manila, Philippines*, pages 100–104, 2005.

7. Richa Agarwala, David L Applegate, Donna Maglott, Gregory D Schuler, and Alejandro A Schäffer. A fast and scalable radiation hybrid map construction and integration strategy. *Genome Research*, 10(3):350–364, 2000.

8. Buthainah Al-Kazemi and Chilukuri K Mohan. Multi-phase generalization of the particle swarm optimization algorithm. In *Proceedings of the 2002 Congress on Evolutionary Computation. CEC'02 (Cat. No. 02TH8600)*, volume 1, pages 489–494. IEEE, 2002.

9. Manish Bachlaus, Mayank Kumar Pandey, Chetan Mahajan, Ravi Shankar, and Manoj Kumar Tiwari. Designing an integrated multi-echelon agile supply chain network: a hybrid taguchi-particle swarm optimization approach. *Journal of Intelligent Manufacturing*, 19(6):747, 2008.

10. Altaf QH Badar, BS Umre, and AS Junghare. Reactive power control using dynamic particle swarm optimization for real power loss minimization. *International Journal of Electrical Power & Energy Systems*, 41(1):133–136, 2012.

11. Altaf QH Badar, BS Umre, and AS Junghare. Reducing variable trend search algorithm for optimizing non linear multidimensional space search. In *2014 International Conference on Advances in Engineering & Technology Research (ICAETR-2014)*, pages 1–4. IEEE, 2014.

12. James E Baker. Reducing bias and inefficiency in the selection algorithm. In *Proceedings of the second international conference on genetic algorithms*, volume 206, pages 14–21, 1987.

13. Javad Behnamian and SMT Fatemi Ghomi. Development of a PSO-SA hybrid metaheuristic for a new comprehensive regression model to time-series forecasting. *Expert Systems with Applications*, 37(2):974–984, 2010.

14. Malika Bessedik, Fatima Benbouzid-Si Tayeb, Hamza Cheurfi, and Ammar Blizak. An immunity-based hybrid genetic algorithms for permutation flowshop scheduling problems. *The International Journal of Advanced Manufacturing Technology*, 85(9-12):2459–2469, 2016.

15. Armando Blanco, Miguel Delgado, and Maria C Pegalajar. A real-coded genetic algorithm for training recurrent neural networks. *Neural Networks*, 14(1):93–105, 2001.

16. Tobias Blickle and Lothar Thiele. A comparison of selection schemes used in evolutionary algorithms. *Evolutionary Computation*, 4(4):361–394, 1996.

17. Janez Brest, Sao Greiner, Borko Boskovic, Marjan Mernik, and Viljem Zumer. Self-adapting control parameters in differential evolution: A comparative study on numerical benchmark problems. *IEEE Transactions on Evolutionary Computation*, 10(6):646–657, 2006.

18. Lance D Chambers. *The practical handbook of genetic algorithms: applications*. Chapman and Hall/CRC, 2000.

19. Jui-Fang Chang, John Francis Roddick, Jeng-Shyang Pan, and Shu-Chuan Chu. A parallel particle swarm optimization algorithm with communication strategies. 2005.

20. Thanga Raj Chelliah, Radha Thangaraj, Srikanth Allamsetty, and Millie Pant. Coordination of directional overcurrent relays using opposition based chaotic differential evolution algorithm. *International Journal of Electrical Power & Energy Systems*, 55:341–350, 2014.

21. W.N. Chen, J. Zhang, Y. Lin, N. Chen, Z.H. Zhan, H.S.H. Chung, Y. Li, and Y.H. Shi. Particle Swarm Optimization with an Aging Leader and Challengers. *Evolutionary Computation, IEEE Transactions on*, 17(2):241–258, April 2013.

22. M. Clerc. The Swarm and the Queen: Towards a Deterministic and Adaptive Particle Swarm Optimization. In *Evolutionary Computation, Proceedings of the 1999 Congress on*, volume 3, pages 1951–1957 Vol. 3, 1999.

23. Ld S Coelho and Viviana Cocco Mariani. Combining of chaotic differential evolution and quadratic programming for economic dispatch optimization with valve-point effect. *IEEE Transactions on Power Systems*, 21(2):989–996, 2006.

24. Matej Črepinšek, Shih-Hsi Liu, and Luka Mernik. A note on teaching–learning-based optimization algorithm. *Information Sciences*, 212:79–93, 2012.

25. Ioannis G Damousis, Anastasios G Bakirtzis, and Petros S Dokopoulos. Network-constrained economic dispatch using real-coded genetic algorithm. *IEEE Transactions on Power Systems*, 18(1):198–205, 2003.

26. Dipankar Dasgupta and Douglas R McGregor. *sGA: A structured genetic algorithm.* Citeseer, 1993.

27. Donald Davendra. *Traveling Salesman Problem: Theory and Applications.* InTech, 2010.

28. Thaís de Fátima Araújo and Wadaed Uturbey. Performance assessment of PSO, DE and hybrid PSO-DE algorithms when applied to the dispatch of generation and demand. *International Journal of Electrical Power & Energy Systems*, 47:205–217, 2013.

29. Kalyanmoy Deb, Ram Bhushan Agrawal, et al. Simulated binary crossover for continuous search space. *Complex Systems*, 9(2):115–148, 1995.

30. Kalyanmoy Deb, Ashish Anand, and Dhiraj Joshi. A computationally efficient evolutionary algorithm for real-parameter optimization. *Evolutionary Computation*, 10(4):371–395, 2002.

31. Kalyanmoy Deb and Hans-Georg Beyer. Self-adaptive genetic algorithms with simulated binary crossover. *Evolutionary Computation*, 9(2):197–221, 2001.

32. Kusum Deep, Krishna Pratap Singh, Mitthan Lal Kansal, and C Mohan. A real coded genetic algorithm for solving integer and mixed integer optimization problems. *Applied Mathematics and Computation*, 212(2):505–518, 2009.

33. Leandro dos Santos Coelho. Reliability–redundancy optimization by means of a chaotic differential evolution approach. *Chaos, Solitons & Fractals*, 41(2):594–602, 2009.

34. Amer Draa, Souham Meshoul, Hichem Talbi, and Mohamed Batouche. A quantum-inspired differential evolution algorithm for solving the N-queens problem. *Neural Networks*, 1(2), 2011.

35. Russ C Eberhart and Yuhui Shi. Comparing Inertia Weights and Constriction Factors in Particle Swarm Optimization. In *Evolutionary Computation, 2000. Proceedings of the 2000 Congress on*, volume 1, pages 84–88. IEEE, 2000.

36. AM Elaiw, X Xia, and AM Shehata. Hybrid DE-SQP and hybrid PSO-SQP methods for solving dynamic economic emission dispatch problem with valve-point effects. *Electric Power Systems Research*, 103:192–200, 2013.

37. Ehab E Elattar. A hybrid genetic algorithm and bacterial foraging approach for dynamic economic dispatch problem. *International Journal of Electrical Power & Energy Systems*, 69:18–26, 2015.

38. Ehab Z Elfeky, Ruhul A Sarker, and Daryl L Essam. Analyzing the simple ranking and selection process for constrained evolutionary optimization. *journal of Computer Science and Technology*, 23(1):19–34, 2008.

39. Saber M Elsayed, Ruhul A Sarker, and Daryl L Essam. A comparative study of different variants of genetic algorithms for constrained optimization. In *Asia-Pacific Conference on Simulated Evolution and Learning*, pages 177–186. Springer, 2010.

40. Larry J Eshelman and J David Schaffer. Real-coded genetic algorithms and interval-schemata. In *Foundations of genetic algorithms*, volume 2, pages 187–202. Elsevier, 1993.

41. Muzaffar Eusuff, Kevin Lansey, and Fayzul Pasha. Shuffled frog-leaping algorithm: a memetic meta-heuristic for discrete optimization. *Engineering Optimization*, 38(2):129–154, 2006.

42. Stefka Fidanova, Marcin Paprzycki, and Olympia Roeva. Hybrid GA-ACO algorithm for a model parameters identification problem. In *2014 Federated Conference on Computer Science and Information Systems*, pages 413–420. IEEE, 2014.

43. B Bahmani Firouzi, M Sha Sadeghi, and Taher Niknam. A new hybrid algorithm based on PSO, SA, and K-means for cluster analysis. *International Journal of Innovative Computing, Information and Control*, 6(7):3177–3192, 2010.

44. Hao Gao, Sam Kwong, Baojie Fan, and Ran Wang. A hybrid particle-swarm tabu search algorithm for solving job shop scheduling problems. *IEEE Transactions on Industrial Informatics*, 10(4):2044–2054, 2014.

45. Harish Garg. A hybrid PSO-GA algorithm for constrained optimization problems. *Applied Mathematics and Computation*, 274:292–305, 2016.

46. Harish Garg. A hybrid GSA-GA algorithm for constrained optimization problems. *Information Sciences*, 478:499–523, 2019.

47. David E Goldberg and Kalyanmoy Deb. A comparative analysis of selection schemes used in genetic algorithms. In *Foundations of genetic algorithms*, volume 1, pages 69–93. Elsevier, 1991.

48. David E Goldberg, Bradley Korb, and Kalyanmoy Deb. Messy genetic algorithms: Motivation, analysis, and first results. *Complex Systems*, 3(5):493–530, 1989.

49. Wenyin Gong, Zhihua Cai, Charles X Ling, and Hui Li. Enhanced differential evolution with adaptive strategies for numerical optimization. *IEEE Transactions on Systems, Man, and Cybernetics, Part B (Cybernetics)*, 41(2):397–413, 2010.

50. John J Greffenstette and James E Baker. How genetic algorithms work: A critical look at implicit parallelism. In *Proceedings of the 3rd International Conference on Genetic Algorithms*, pages 20–27, 1989.

51. Jinwei Gu, Manzhan Gu, Cuiwen Cao, and Xingsheng Gu. A novel competitive co-evolutionary quantum genetic algorithm for stochastic job shop scheduling problem. *Computers & Operations Research*, 37(5):927–937, 2010.

52. Gregory Gutin and Abraham P Punnen. *The traveling salesman problem and its variations*, volume 12. Springer Science & Business Media, 2006.

53. Peter JB Hancock. A comparison of selection mechanisms. *Bäck et al. [3]*, pages 212–227, 2000.

54. Dakuo He, Fuli Wang, and Zhizhong Mao. A hybrid genetic algorithm approach based on differential evolution for economic dispatch with valve-point effect. *International Journal of Electrical Power & Energy Systems*, 30(1):31–38, 2008.

55. Nicholas Holden and Alex A Freitas. Hierarchical classification of g-protein-coupled receptors with a PSO/ACO algorithm. In *Proceedings of the IEEE Swarm Intelligence Symposium (SIS'06)*, pages 77–84. IEEE Press, 2006.

56. Nicholas Holden and Alex A Freitas. A hybrid PSO/ACO algorithm for discovering classification rules in data mining. *Journal of Artificial Evolution and Applications*, 2008.

57. Nicholas Paul Holden and Alex A Freitas. A hybrid PSO/ACO algorithm for classification. In *Proceedings of the 9th annual conference companion on Genetic and evolutionary computation*, pages 2745–2750, 2007.

58. Lhassane Idoumghar, Mahmoud Melkemi, René Schott, and Maha Idrissi Aouad. Hybrid PSO-SA type algorithms for multimodal function optimization and reducing energy consumption in embedded systems. *Applied Computational Intelligence and Soft Computing*, 2011, 2011.

59. Muhammad Amjad Iqbal, Naveed Kazim Khan, Hasan Mujtaba, and A Rauf Baig. A novel function optimization approach using opposition based genetic algorithm with gene excitation. *International Journal of Innovative Computing, Information and Control*, 7(7), 2011.

60. Shimpi Singh Jadon, Ritu Tiwari, Harish Sharma, and Jagdish Chand Bansal. Hybrid artificial bee colony algorithm with differential evolution. *Applied Soft Computing*, 58:11–24, 2017.

61. Suhanya Jayaprakasam, Sharul Kamal Abdul Rahim, and Chee Yen Leow. PSOGSA - explore: A new hybrid metaheuristic approach for beampattern optimization in collaborative beamforming. *Applied Soft Computing*, 30:229–237, 2015.

62. Alison Jenkins, Vinika Gupta, Alexis Myrick, and Mary Lenoir. Variations of genetic algorithms. *arXiv preprint arXiv:1911.00490*, 2019.

63. Wei Jiang, Kai-Qing Zhou, and Li-Ping Mo. Parameter optimization strategy of fuzzy petri net utilizing hybrid GA-SFLA algorithm. In *International Conference on Simulation Tools and Techniques*, pages 416–426. Springer, 2019.

64. Dervis Karaboga. An idea based on honey bee swarm for numerical optimization. Technical report, Technical report-tr06, Erciyes university, engineering faculty, computer ..., 2005.

65. Dervis Karaboga and Bahriye Akay. A comparative study of artificial bee colony algorithm. *Applied Mathematics and Computation*, 214(1):108–132, 2009.

66. S Kayalvili and M Selvam. Hybrid SFLA-GA algorithm for an optimal resource allocation in cloud. *Cluster Computing*, 22(2):3165–3173, 2019.

67. J. Kennedy and R. Eberhart. Particle Swarm Optimization. In *Neural Networks, 1995. Proceedings., IEEE International Conference on*, volume 4, pages 1942–1948, November 1995.

68. James Kennedy. The Particle Swarm: Social Adaptation of Knowledge. In *Evolutionary Computation, IEEE International Conference on*, pages 303–308. IEEE, 1997.

69. Rajendra Ku Khadanga and Jitendriya Ku Satapathy. A new hybrid GA-GSA algorithm for tuning damping controller parameters for a unified power flow controller. *International Journal of Electrical Power & Energy Systems*, 73:1060–1069, 2015.

70. S Khamsawang, P Wannakarn, and S Jiriwibhakorn. Hybrid PSO-DE for solving the economic dispatch problem with generator constraints. In *2010 the 2nd international conference on computer and automation engineering (ICCAE)*, volume 5, pages 135–139. IEEE, 2010.

71. Thiemo Krink, Jakob S VesterstrOm, and Jacques Riget. Particle swarm optimisation with spatial particle extension. In *Proceedings of the 2002 Congress on Evolutionary Computation. CEC'02 (Cat. No. 02TH8600)*, volume 2, pages 1474–1479. IEEE, 2002.

72. C Kumar, S Prakash, T Kumar, and DP Sahu. Variant of genetic algorithm and its applications. In *International Conference on Advances in Computer and Electronics Technology, Hong Kong*, pages 25–29, 2014.

73. Sajjan Kumar, Kamal K Mandal, and Niladri Chakraborty. Optimal DG placement by multi-objective opposition based chaotic differential evolution for techno-economic analysis. *Applied Soft Computing*, 78:70–83, 2019.

74. Eugene L Lawler, Jan Karel Lenstra, Alexander HG Rinnooy Kan, and David Bernard Shmoys. The traveling salesman problem; a guided tour of combinatorial optimization. 1985.

75. Jie Li, Ding Fan, and Victor Sreeram. Sfc optimization for aero engine based on hybrid GA-SQP method. *International Journal of Turbo & Jet-engines*, 30(4):383–391, 2013.

76. Xiangtao Li and Minghao Yin. Parameter estimation for chaotic systems by hybrid differential evolution algorithm and artificial bee colony algorithm. *Nonlinear Dynamics*, 77(1–2):61–71, 2014.

77. Xiangtao Li, Minghao Yin, and Zhiqiang Ma. Hybrid differential evolution and gravitation search algorithm for unconstrained optimization. *International Journal of Physical Sciences*, 6(25):5961–5981, 2011.

78. Yuancheng Li, Yiliang Wang, and Bin Li. A hybrid artificial bee colony assisted differential evolution algorithm for optimal reactive power flow. *International Journal of Electrical Power & Energy Systems*, 52:25–33, 2013.

79. Zhiyong Li, Weiyou Wang, Yanyan Yan, and Zheng Li. PS–ABC: A hybrid algorithm based on particle swarm and artificial bee colony for high-dimensional optimization problems. *Expert Systems with Applications*, 42(22):8881–8895, 2015.

80. Jing J Liang, A Kai Qin, Ponnuthurai N Suganthan, and S Baskar. Comprehensive learning particle swarm optimizer for global optimization of multimodal functions. *IEEE Transactions on Evolutionary Computation*, 10(3):281–295, 2006.

81. Qiuzhen Lin, Qingling Zhu, Peizhi Huang, Jianyong Chen, Zhong Ming, and Jianping Yu. A novel hybrid multi-objective immune algorithm with adaptive differential evolution. *Computers & Operations Research*, 62:95–111, 2015.

82. Wen-Yi Lin. A GA–DE hybrid evolutionary algorithm for path synthesis of four-bar linkage. *Mechanism and Machine Theory*, 45(8):1096–1107, 2010.

83. Bo Liu, Ling Wang, Yi-Hui Jin, Fang Tang, and De-Xian Huang. Improved particle swarm optimization combined with chaos. *Chaos, Solitons & Fractals*, 25(5):1261–1271, 2005.

84. Junhong Liu and Jouni Lampinen. A fuzzy adaptive differential evolution algorithm. *Soft Computing*, 9(6):448–462, 2005.

85. Yubao Liu and Guihe Qin. A hybrid TS-DE algorithm for reliability redundancy optimization problem. *Journal of Computers*, 9(9):2050–7, 2014.

86. Youlin Lu, Jianzhong Zhou, Hui Qin, Ying Wang, and Yongchuan Zhang. An adaptive chaotic differential evolution for the short-term hydrothermal generation scheduling problem. *Energy Conversion and Management*, 51(7):1481–1490, 2010.

87. Youlin Lu, Jianzhong Zhou, Hui Qin, Ying Wang, and Yongchuan Zhang. Chaotic differential evolution methods for dynamic economic dispatch with valve-point effects. *Engineering Applications of Artificial Intelligence*, 24(2):378–387, 2011.

88. Esmaeil Mahboubi-Moghaddam, Jamshid Aghaei, Kashem M Muttaqi, Behrouz Zoghdar-Moghadam-Shahrekohne, and Mohammad Rasoul Narimani. Evaluating distributed generations in utility operation and planning issues using a novel fusion PSO-SFLA algorithm. In *2015 Australasian Universities Power Engineering Conference (AUPEC)*, pages 1–6. IEEE, 2015.

89. Isa Maleki, Ali Ghaffari, and Mohammad Masdari. A new approach for software cost estimation with hybrid genetic algorithm and ant colony optimization. *International Journal of Innovation and Applied Studies*, 5(1):72, 2014.

90. Rammohan Mallipeddi, Ponnuthurai N Suganthan, Quan-Ke Pan, and Mehmet Fatih Tasgetiren. Differential evolution algorithm with ensemble of parameters and mutation strategies. *Applied Soft Computing*, 11(2):1679–1696, 2011.

91. Behrang Mansoornejad, Navid Mostoufi, and Farhang Jalali-Farahani. A hybrid ga–sqp optimization technique for determination of kinetic parameters of hydrogenation reactions. *Computers & Chemical Engineering*, 32(7):1447–1455, 2008.

92. Rui Mendes, James Kennedy, and Josè Neves. The fully informed particle swarm: simpler, maybe better. *IEEE Transactions on Evolutionary Computation*, 8(3):204–210, 2004.

93. Efrñn Mezura-Montes, Jesús Velázquez-Reyes, and Carlos A Coello Coello. A comparative study of differential evolution variants for global optimization. In *Proceedings of the 8th Annual Conference on Genetic and Evolutionary Computation*, pages 485–492, 2006.

94. Said M Mikki and Ahmed A Kishk. Quantum particle swarm optimization for electromagnetics. *IEEE Transactions on Antennas and Propagation*, 54(10):2764–2775, 2006.

95. Byoung-Mun Min, Hyeok Ryu, Daekyu Sang, Min-Jea Tahk, and David Hyunchul Shim. Autopilot design using hybrid PSO-SQP algorithm. In *International Conference on Intelligent Computing*, pages 596–604. Springer, 2007.

96. Seyedali Mirjalili and Siti Zaiton Mohd Hashim. A new hybrid PSOGSA algorithm for function optimization. In *2010 International Conference on Computer and Information Application*, pages 374–377. IEEE, 2010.

97. SeyedAli Mirjalili, Siti Zaiton Mohd Hashim, and Hossein Moradian Sardroudi. Training feedforward neural networks using hybrid particle swarm optimization and gravitational search algorithm. *Applied Mathematics and Computation*, 218(22):11125–11137, 2012.

98. Seyedali Mirjalili and Andrew Lewis. The whale optimization algorithm. *Advances in Engineering Software*, 95:51–67, 2016.

99. Seyedali Mirjalili, Seyed Mohammad Mirjalili, and Andrew Lewis. Grey wolf optimizer. *Advances in Engineering Software*, 69:46–61, 2014.

100. Heinz Mühlenbein, M Schomisch, and Joachim Born. The parallel genetic algorithm as function optimizer. *Parallel Computing*, 17(6–7):619–632, 1991.

101. Tadahiko Murata and Hisao Ishibuchi. MOGA: multi-objective genetic algorithms. In *IEEE International Conference on Evolutionary Computation*, volume 1, pages 289–294, 1995.

102. Taher Niknam, Babak Amiri, Javad Olamaei, and Ali Arefi. An efficient hybrid evolutionary optimization algorithm based on PSO and SA for clustering. *Journal of Zhejiang University-SCIENCE A*, 10(4):512–519, 2009.

103. Ben Niu and Li Li. A novel PSO-DE-based hybrid algorithm for global optimization. In *International Conference on Intelligent Computing*, pages 156–163. Springer, 2008.

104. Mahamed GH Omran, Andries P Engelbrecht, and Ayed Salman. Bare bones differential evolution. *European Journal of Operational Research*, 196(1):128–139, 2009.

105. Isao Ono, Hajime Kita, and Shigenobu Kobayashi. A real-coded genetic algorithm using the unimodal normal distribution crossover. In *Advances in Evolutionary Computing*, pages 213–237. Springer, 2003.

106. H Orouji, O Bozorg Haddad, E Fallah-Mehdipour, and MA Mariño. Extraction of decision alternatives in project management: Application of hybrid PSO-SFLA. *Journal of Management in Engineering*, 30(1):50–59, 2014.

107. Gary Pampara, Andries Petrus Engelbrecht, and Nelis Franken. Binary differential evolution. In *2006 IEEE International Conference on Evolutionary Computation*, pages 1873–1879. IEEE, 2006.

108. Feng Pan, Xiaohui Hu, Russ Eberhart, and Yaobin Chen. An analysis of bare bones particle swarm. In *2008 IEEE Swarm Intelligence Symposium*, pages 1–5. IEEE, 2008.

109. Millie Pant, Radha Thangaraj, and Ajith Abraham. DE-PSO: a new hybrid metaheuristic for solving global optimization problems. *New Mathematics and Natural Computation*, 7(03):363–381, 2011.

110. Konstantinos E Parsopoulos and Michael N Vrahatis. Unified particle swarm optimization in dynamic environments. In *Workshops on Applications of Evolutionary Computation*, pages 590–599. Springer, 2005.

111. Kevin M Passino. Biomimicry of bacterial foraging for distributed optimization and control. *IEEE Control Systems Magazine*, 22(3):52–67, 2002.

112. Magnus Erik Hvass Pedersen. Good Parameters for Particle Swarm Optimization. *Hvass Lab., Copenhagen, Denmark, Tech. Rep. HL1001*, 2010.

113. Magnus Erik Hvass Pedersen and Andrew J Chipperfield. Simplifying Particle Swarm Optimization. *Applied Soft Computing*, 10(2):618–628, 2010.

114. Chunhua Peng, Huijuan Sun, Jianfeng Guo, and Gang Liu. Dynamic economic dispatch for wind-thermal power system using a novel bi-population chaotic differential evolution algorithm. *International Journal of Electrical Power & Energy Systems*, 42(1):119–126, 2012.

115. Thanmaya Peram, Kalyan Veeramachaneni, and Chilukuri K Mohan. Fitness-distance-ratio based particle swarm optimization. In *Proceedings of the 2003 IEEE Swarm Intelligence Symposium. SIS'03 (Cat. No. 03EX706)*, pages 174–181. IEEE, 2003.

116. Yiannis G Petalas, Konstantinos E Parsopoulos, and Michael N Vrahatis. Memetic particle swarm optimization. *Annals of Operations Research*, 156(1):99–127, 2007.

117. Michal Pluhacek, Roman Senkerik, Donald Davendra, Zuzana Kominkova Oplatkova, and Ivan Zelinka. On the behavior and performance of chaos driven PSO algorithm with inertia weight. *Computers & Mathematics with Applications*, 66(2):122–134, 2013.

118. Michal Pluhacek, Roman Senkerik, Ivan Zelinka, and Donald Davendra. Chaos PSO algorithm driven alternately by two different chaotic maps – an initial study. In *2013 IEEE Congress on Evolutionary Computation*, pages 2444–2449. IEEE, 2013.

119. Sweta Potthuri, T Shankar, and A Rajesh. Lifetime improvement in wireless sensor networks using hybrid differential evolution and simulated annealing (DESA). *Ain Shams Engineering Journal*, 9(4):655–663, 2018.

120. Prasenjit Pramanik and Manas Kumar Maiti. An inventory model for deteriorating items with inflation induced variable demand under two level partial trade credit: A hybrid ABC-GA approach. *Engineering Applications of Artificial Intelligence*, 85:194–207, 2019.

121. K Premalatha and AM Natarajan. Hybrid PSO and GA for global maximization. *International Journal of Open Problems in Computer Science and Mathematics*, 2(4):597–608, 2009.

122. Constantin Purcaru, Radu-Emil Precup, Daniel Iercan, Lucian-Ovidiu Fedorovici, and Radu-Codrut David. Hybrid PSO-GSA robot path planning algorithm in static environments with danger zones. In *2013 17th International Conference on System Theory, Control and Computing (ICSTCC)*, pages 434–439. IEEE, 2013.

123. A Kai Qin, Vicky Ling Huang, and Ponnuthurai N Suganthan. Differential evolution algorithm with strategy adaptation for global numerical optimization. *IEEE Transactions on Evolutionary Computation*, 13(2):398–417, 2008.

124. Nur Azzammudin Rahmat and I Musirin. Differential evolution ant colony optimization (DEACO) technique in solving economic load dispatch problem. In *2012 IEEE International Power Engineering and Optimization Conference Melaka, Malaysia*, pages 263–268. IEEE, 2012.

125. Shahryar Rahnamayan, Hamid R Tizhoosh, and Magdy MA Salama. Opposition-based differential evolution algorithms. In *2006 IEEE International Conference on Evolutionary Computation*, pages 2010–2017. IEEE, 2006.

126. R Venkata Rao. Teaching-learning-based optimization algorithm. In *Teaching learning based optimization algorithm*, pages 9–39. Springer, 2016.

127. R Venkata Rao, Vimal J Savsani, and DP Vakharia. Teaching–learning-based optimization: a novel method for constrained mechanical design optimization problems. *Computer-Aided Design*, 43(3):303–315, 2011.

128. R Venkata Rao, Vimal J Savsani, and DP Vakharia. Teaching–learning-based optimization: an optimization method for continuous non-linear large scale problems. *Information Sciences*, 183(1):1–15, 2012.

129. Esmat Rashedi, Hossein Nezamabadi-Pour, and Saeid Saryazdi. GSA: a gravitational search algorithm. *Information sciences*, 179(13):2232–2248, 2009.

130. Abdelmadjid Recioui and Hamid Bentarzi. Capacity optimization of MIMO wireless communication systems using a hybrid genetic-taguchi algorithm. *Wireless Personal Communications*, 71(2):1003–1019, 2013.

131. S Surender Reddy. Optimal power flow using hybrid differential evolution and harmony search algorithm. *International Journal of Machine Learning and Cybernetics*, 10(5):1077–1091, 2019.

132. Jacques Riget and Jakob S Vesterstrøm. A diversity-guided particle swarm optimizer-the ARPSO. *Dept. Comput. Sci., Univ. of Aarhus, Aarhus, Denmark, Tech. Rep*, 2:1–13, 2002.

133. Tea Robič and Bogdan Filipič. Differential evolution for multiobjective optimization. In *International Conference on Evolutionary Multi-Criterion Optimization*, pages 520–533. Springer, 2005.

134. Zahir Sahli, Abdelatif Hamouda, Abdelghani Bekrar, and Damien Trentesaux. Hybrid PSO-tabu search for the optimal reactive power dispatch problem. In *IECON 2014-40th Annual Conference of the IEEE Industrial Electronics Society*, pages 3536–3542. IEEE, 2014.

135. Omar M Sallabi and Younis R Elhaddad. A new hybrid genetic and simulated annealing algorithm to solve the traveling salesman problem. 2014.

136. Jeffrey R Sampson. Adaptation in natural and artificial systems (John H. Holland), 1976.

137. Alexander Schrijver. On the history of combinatorial optimization (until 1960). *Handbooks in operations research and management science*, 12:1–68. Elsevier, 2005.

138. Mourad Sefrioui and Jacques Périaux. A hierarchical genetic algorithm using multiple models for optimization. In *International Conference on Parallel Problem Solving From Nature*, pages 879–888. Springer, 2000.

139. Qi Shen, Wei-Min Shi, and Wei Kong. Hybrid particle swarm optimization and tabu search approach for selecting genes for tumor classification using gene expression data. *Computational Biology and Chemistry*, 32(1):53–60, 2008.

140. XH Shi, YC Liang, HP Lee, Chun Lu, and LM Wang. An improved GA and a novel PSO-GA-based hybrid algorithm. *Information Processing Letters*, 93(5):255–261, 2005.

141. Yuhui Shi and R.C. Eberhart. Empirical Study of Particle Swarm Optimization. In *Evolutionary Computation, 1999. CEC 99. Proceedings of the 1999 Congress on*, volume 3, pages –1950, 1999.

142. Yuhui Shi and Russell Eberhart. A Modified Particle Swarm Optimizer. In *Evolutionary Computation Proceedings, 1998. IEEE World Congress on Computational Intelligence., The 1998 IEEE International Conference on*, pages 69–73. IEEE, 1998.

143. Yuhui Shi and Russell C Eberhart. Parameter Selection in Particle Swarm Optimization. In *Evolutionary Programming VII*, pages 591–600. Springer, 1998.

144. Margarita Sordo, Gabriela Ochoa, and Shawn N Murphy. A PSO/ACO approach to knowledge discovery in a pharmacovigilance context. In *Proceedings of the 11th Annual Conference Companion on Genetic and Evolutionary Computation Conference: Late Breaking Papers*, pages 2679–2684, 2009.

145. Mandavilli Srinivas and Lalit M Patnaik. Adaptive probabilities of crossover and mutation in genetic algorithms. *IEEE Transactions on Systems, Man, and Cybernetics*, 24(4):656–667, 1994.

146. Rainer Storn and Kenneth Price. *Differential Evolution - A Simple and Efficient Adaptive Scheme for Global Optimization over Continuous Spaces*. ICSI Berkeley, 1995.

147. Rainer Storn and Kenneth Price. Differential evolution–a simple and efficient heuristic for global optimization over continuous spaces. *Journal of Global Optimization*, 11(4):341–359, 1997.

148. Jun Sun, Xiaojun Wu, Vasile Palade, Wei Fang, and Yuhui Shi. Random Drift Particle Swarm Optimization. *arXiv preprint arXiv:1306.2863*, 2013.

149. Xun SUN, WG Zhang, Wei Yin, and AJ Li. Optimization of flight controller parameters based on PSO-immune algorithm. *Journal of System Simulation*, 19(12):2765–2767, 2007.

150. M Fatih Taşgetiren and Yun-Chia Liang. A binary particle swarm optimization algorithm for lot sizing problem. *Journal of Economic and Social Research*, 5(2):1–20, 2003.

151. Valery Tereshko. Reaction-diffusion model of a honeybee colony's foraging behavior. In *International Conference on Parallel Problem Solving from Nature*, pages 807–816. Springer, 2000.

152. Valery Tereshko and Troy Lee. How information-mapping patterns determine foraging behavior of a honey bee colony. *Open Systems & Information Dynamics*, 9(02):181–193, 2002.

153. Valery Tereshko and Andreas Loengarov. Collective decision making in honey-bee foraging dynamics. *Computing and Information Systems*, 9(3):1, 2005.

154. TO Ting, MVC Rao, and CK Loo. A novel approach for unit commitment problem via an effective hybrid particle swarm optimization. *IEEE Transactions on Power Systems*, 21(1):411–418, 2006.

155. Ville Tirronen, Ferrante Neri, Tommi Karkkainen, Kirsi Majava, and Tuomo Rossi. A memetic differential evolution in filter design for defect detection in paper production. In *Workshops on Applications of Evolutionary Computation*, pages 320–329. Springer, 2007.

156. S Titus and A Ebenezer Jeyakumar. A hybrid EP-PSO-SQP algorithm for dynamic dispatch considering prohibited operating zones. *Electric power components and systems*, 36(5):449–467, 2008.

157. Duc-Hoc Tran, Min-Yuan Cheng, and Minh-Tu Cao. Hybrid multiple objective artificial bee colony with differential evolution for the time–cost–quality tradeoff problem. *Knowledge-Based Systems*, 74:176–186, 2015.

158. Anupam Trivedi, Krishnendu Sanyal, Pranjal Verma, and Dipti Srinivasan. A unified differential evolution algorithm for constrained optimization problems. In *2017 IEEE congress on evolutionary computation (CEC)*, pages 1231–1238. IEEE, 2017.

159. Jinn-Tsong Tsai, Jyh-Horng Chou, and Tung-Kuan Liu. Tuning the structure and parameters of a neural network by using hybrid taguchi-genetic algorithm. *IEEE Transactions on Neural Networks*, 17(1):69–80, 2006.

160. Jinn-Tsong Tsai, Tung-Kuan Liu, and Jyh-Horng Chou. Hybrid taguchi-genetic algorithm for global numerical optimization. *IEEE Transactions on Evolutionary Computation*, 8(4):365–377, 2004.

161. Shigeyoshi Tsutsui, Masayuki Yamamura, and Takahide Higuchi. Multi-parent recombination with simplex crossover in real coded genetic algorithms. In *Proceedings of the 1st Annual Conference on Genetic and Evolutionary Computation, volume 1*, pages 657–664, 1999.

162. Frans van den Bergh and Andries P Engelbrecht. A new locally convergent particle swarm optimiser. In *IEEE International Conference on Systems, Man and Cybernetics*, volume 3, pages 6–pp. IEEE, 2002.

163. Frans Van den Bergh and Andries Petrus Engelbrecht. A cooperative approach to particle swarm optimization. *IEEE Transactions on Evolutionary Computation*, 8(3):225–239, 2004.

164. T Aruldoss Albert Victoire and A Ebenezer Jeyakumar. Hybrid PSO–SQP for economic dispatch with valve-point effect. *Electric Power Systems Research*, 71(1):51–59, 2004.

165. Dongshu Wang, Dapei Tan, and Lei Liu. Particle swarm optimization algorithm: an overview. *Soft Computing*, 22(2):387–408, 2018.

166. Hui Wang, Hui Li, Yong Liu, Changhe Li, and Sanyou Zeng. Opposition-based particle swarm algorithm with cauchy mutation. In *2007 IEEE Congress on Evolutionary Computation*, pages 4750–4756. IEEE, 2007.

167. Jiahai Wang, Weiwei Zhang, and Jun Zhang. Cooperative differential evolution with multiple populations for multiobjective optimization. *IEEE Transactions on Cybernetics*, 46(12):2848–2861, 2015.

168. Shuihua Wang, Yudong Zhang, Zhengchao Dong, Sidan Du, Genlin Ji, Jie Yan, Jiquan Yang, Qiong Wang, Chunmei Feng, and Preetha Phillips. Feed-forward neural network optimized by hybridization of PSO and ABC for abnormal brain detection. *International Journal of Imaging Systems and Technology*, 25(2):153–164, 2015.

169. Yong Wang, Zixing Cai, and Qingfu Zhang. Differential evolution with composite trial vector generation strategies and control parameters. *IEEE Transactions on Evolutionary Computation*, 15(1):55–66, 2011.

170. Yuanlong Wang, Wanzhong Zhao, Guan Zhou, Qiang Gao, and Chunyan Wang. Optimization of an auxetic jounce bumper based on gaussian process metamodel and series hybrid GA-SQP algorithm. *Structural and Multidisciplinary Optimization*, 57(6):2515–2525, 2018.

171. Alden H Wright. Genetic algorithms for real parameter optimization. In *Foundations of genetic algorithms*, volume 1, pages 205–218. Elsevier, 1991.

172. Guohua Wu, Xin Shen, Haifeng Li, Huangke Chen, Anping Lin, and Ponnuthurai N Suganthan. Ensemble of differential evolution variants. *Information Sciences*, 423:172–186, 2018.

173. Fatos Xhafa, Juan A Gonzalez, Keshav P Dahal, and Ajith Abraham. A GA(TS) hybrid algorithm for scheduling in computational grids. In *International Conference on Hybrid Artificial Intelligence Systems*, pages 285–292. Springer, 2009.

174. Wanli Xiang, Shoufeng Ma, and Meiqing An. Habcde: a hybrid evolutionary algorithm based on artificial bee colony algorithm and differential evolution. *Applied Mathematics and Computation*, 238:370–386, 2014.

175. Xiaolan Xie, Ruikun Liu, Xiaochun Cheng, Xin Hu, and Jinsheng Ni. Trust-driven and PSO-SFLA based job scheduling algorithm on cloud. *Intelligent Automation & Soft Computing*, 22(4):561–566, 2016.

176. Xin-She Yang. *Nature-inspired metaheuristic algorithms*. Luniver Press, 2010.

177. Xin-She Yang. A new metaheuristic bat-inspired algorithm. In *Nature inspired cooperative strategies for optimization (NICSO 2010)*, pages 65–74. Springer, 2010.

178. Zhenyu Yang, Ke Tang, and Xin Yao. Self-adaptive differential evolution with neighborhood search. In *2008 IEEE Congress on Evolutionary Computation (IEEE World Congress on Computational Intelligence)*, pages 1110–1116. IEEE, 2008.

179. Firas Yengui, Lioua Labrak, Felipe Frantz, Renaud Daviot, Nacer Abouchi, and Ian O'Connor. A hybrid GA-SQP algorithm for analog circuits sizing. 2012.

180. Ali R Yildiz. Hybrid taguchi-differential evolution algorithm for optimization of multipass turning operations. *Applied Soft Computing*, 13(3):1433–1439, 2013.

181. Shiwei Yu, Chang Ding, and Kejun Zhu. A hybrid GA-TS algorithm for open vehicle routing optimization of coal mines material. *Expert Systems with Applications*, 38(8):10568–10573, 2011.

182. Shiwei Yu, Yi-Ming Wei, and Ke Wang. A PSO-GA optimal model to estimate primary energy demand of China. *Energy Policy*, 42:329–340, 2012.

183. Xiaohui Yuan, Yanbin Yuan, and Yongchuan Zhang. A hybrid chaotic genetic algorithm for short-term hydro system scheduling. *Mathematics and Computers in Simulation*, 59(4):319–327, 2002.

184. Guohui Zhang, Xinyu Shao, Peigen Li, and Liang Gao. An effective hybrid particle swarm optimization algorithm for multi-objective flexible job-shop scheduling problem. *Computers & Industrial Engineering*, 56(4):1309–1318, 2009.

185. Jingqiao Zhang and Arthur C Sanderson. Jade: adaptive differential evolution with optional external archive. *IEEE Transactions on Evolutionary Computation*, 13(5):945–958, 2009.

186. Jun Zhang, Henry Shu-Hung Chung, and Wai-Lun Lo. Clustering-based adaptive crossover and mutation probabilities for genetic algorithms. *IEEE Transactions on Evolutionary Computation*, 11(3):326–335, 2007.

187. Wen Zhang, Yutian Liu, and Maurice Clerc. An Adaptive PSO Algorithm for Reactive Power Optimization. *6th International Conference on Advances in Power System Control, Operation and Management (APSCOM 2003)*, pages 302–307, November 2003.

188. Yudong Zhang, Yan Jun, Geng Wei, and Lenan Wu. Find multi-objective paths in stochastic networks via chaotic immune PSO. *Expert Systems with Applications*, 37(3):1911–1919, 2010.

189. Guopeng Zhu, Ting Jiang, and Zheng Zhou. A fusion method for object identification in rainy weather based on PSO-SFLA and UWB. In *2016 16th International Symposium on Communications and Information Technologies (ISCIT)*, pages 537–541. IEEE, 2016.

190. Ezgi Zorarpacı and Selma Ayşe Özel. A hybrid approach of differential evolution and artificial bee colony for feature selection. *Expert Systems with Applications*, 62:91–103, 2016.

Index